**Food Science and Technology**

nova
Medicine & Health
New York

# Food Science and Technology

**Food Byproducts: Valorization Through Nutraceutical Production**
Sandra Teresita Martín-del-Campo, PhD (Editor)
Anaberta Cardador-Martínez, PhD (Editor)
Jessica del Pilar Ramírez-Anaya, PhD (Editor)
2023. ISBN: 979-8-88697-612-0 (Hardcover)
2023. ISBN: 979-8-88697-876-6 (eBook)

**Bioprospecting of Natural Compounds in Food, Pharmaceutical and Biomedical Science**
Siddhartha Pati, PhD (Editor)
Tanmay Sarkar, PhD (Editor)
Dibyait Lahiri, PhD (Editor)
2022. ISBN: 979-8-88697-361-7 (Hardcover)
2022. ISBN: 979-8-88697-431-7 (eBook)

**Proceedings of BIOSPECTRUM: The International Conference on Biotechnology and Biological Sciences: Biotechnological Intervention Towards Enhancing Food Value**
Sanket Joshi, PhD (Editor)
Susmita Mukherjee (Editor)
Moupriya Nag, PhD (Editor)
2022. ISBN: 978-1-68507-985-7 (Hardcover)
2022. ISBN: 979-8-88697-069-2 (eBook)

**Nutraceuticals: Food Applications and Health Benefits**
Anita Kumari, PhD (Editor)
Gulab Singh, PhD (Editor)
2022. ISBN: 978-1-68507-488-3 (Hardcover)
2022. ISBN: 978-1-68507-512-5 (eBook)

More information about this series can be found at
https://novapublishers.com/product-category/series/food-science-and-technology-series/

**Daniele de Almeida Paula**
**Karen Cristina Guedes Silva**
**and Nataly de Almeida Costa**
Editors

# Microencapsulation of Probiotics

## Challenges and Future Prospects

nova
Medicine & Health
New York

**Library of Congress Cataloging-in-Publication Data**

ISBN: 979-8-88697-979-4

*Published by Nova Science Publishers, Inc. † New York*

# Contents

# Preface

Probiotics are live microorganisms known to promote health benefits to the host when ingested regularly. Such benefits are dependent on the administrated dose, probiotic viability and stability in the food matrix, survival during the shelf life of added food product, and in the gastrointestinal tract environment. Evidence presented by the literature has demonstrated the action of probiotics in the proper functioning of the gastrointestinal system, increasing immunity, preventing allergies, reducing cholesterol, and combating diabetes, and obesity, among others.

Due to so many potentialities, such microorganisms have been applied in different food matrices such as dairy, meat, and vegetables to promote different alternatives of probiotic foods to the consumer. However, since probiotics are sensitive to some process conditions (such as pH, ionic strength, temperature, oxygen, and chelation agents) and the conditions they are exposed to during consumption (enzymes, stomach acidity) before reaching the site of action (intestine), encapsulation techniques have been extensively studied to provide protection, greater viability and controlled delivery of probiotics.

Microencapsulation has been known as a process used to protect bioactive compounds inside polymeric matrices. The technique is used for different purposes such as in the protection of the core material from adverse conditions, in the promotion of controlled delivery, and for preserving and/or masking of flavors. Currently, several techniques have been used to produce encapsulation systems, but to achieve mastery of these techniques, pioneering studies were fundamental for the development of the microencapsulation area by science. Therefore, this book addresses in one of its first chapters a brief

history of the development of encapsulation systems and their valuable contributions to the various techniques that are currently applied to protect bioactives such as probiotics, as well as the main wall materials used for the production of encapsulation systems. For probiotic application in food, many biopolymers have been studied, these substances must be of food quality, biodegradable, low cost, capable of acting as a physical barrier, protecting the probiotics from the external environment, in addition to promoting the release of the nucleus at the site of action. For these purposes, the use of carbohydrates, proteins, and lipids can be mentioned.

In the following chapters, some encapsulation techniques are presented, highlighting process conditions, materials used, characteristics of the encapsulation systems for probiotic protection, challenges, advantages, and disadvantages. Among the various techniques, a special look was directed at the techniques of ionic gelation due to its simplicity and low cost, in addition to mild process conditions; spray drying, an efficient and reproducible technique on an industrial scale; emulsification, process that also presents mild processing conditions, allowing to maintain the integrity and viability of the probiotic cells and complex coacervation, an emerging microencapsulation technique for probiotic cells considered advantageous because it does not use organic solvents or high temperatures. The combination of encapsulation techniques to improve probiotic protection, stability, and controlled delivery, was also explored in this book.

In sequence, coating layers are approached as an alternative to increasing the protection of encapsulation systems for the protection of probiotics. Coating layers can largely impact the protection of probiotics, as they can help to overcome the weakness of single-layer particles such as the high porosity and permeability, and poor mucoadhesiveness. Depending on the coating and/or core material composition, good protection of the probiotics can be achieved, improving the defense of cells against adverse conditions.

Finally, in vitro and in vivo digestion, as well as the different models used to evaluate the efficiency of microencapsulation techniques and the survival of probiotics, were described. This approach allows presenting information

related to the application of probiotics in food matrices for delivery at the appropriate site of action and in sufficient quantities to promote the intended benefits.

The great development that science has provided in the area of encapsulation is notorious. Currently, the variety of encapsulation systems, from the simplest to the sophisticated systems, in addition to the range of materials and processes, ranging from simple to complex equipment, allow the production of countless particles. In addition to improving protection, and ensuring controlled delivery, one of the main driving forces for research development focuses on using renewable, biocompatible, more environmentally friendly, and safe-to-consume materials. In this sense, it is worth mentioning the importance of research for scaling process that allow industrial production and cost reduction to increase the supply of encapsulated probiotic products in the food market. The evaluation of the particle's interaction with the food matrix to determine their influence on the physical, nutritional, and sensory properties of the foods to which they are applied is also of great importance.

*Daniele de Almeida Paula,* PhD
Professor, Dr., Federal Institute of Paraná, IFPR, Brazil

*Karen Cristina Guedes Silva,* PhD
Profofessor, Dr., Federal Institute of Paraná, IFPR, Brazil

*Nataly de Almeida Costa,* MSc.
Federal University of Viçosa, UFV, Brazil

# Chapter 1

# Probiotics

**Vanessa Caroline de Oliveira***
**Nataly de Almeida Costa**
**and Maria José do Amaral e Paiva**
Department of Food Technology (DTA), Federal University of Viçosa (UFV), Viçosa, Minas Gerais, Brazil

## Abstract

Probiotics are live microorganisms that, when ingested regularly and in sufficient quantities, are capable of promoting benefits, including helping the gastrointestinal system to function properly, increasing immunity, and preventing allergies, among others. Such microorganisms have been applied in different food matrices such as dairy, meat, and vegetables to promote different alternatives of probiotic foods to the consumer. However, some characteristics of the food matrix can affect the viability of these microorganisms and, thus, influence the delivery at the site of action in sufficient quantities to promote the benefits. In this chapter, a brief history of the first reports of the use of probiotic microorganisms and their current definition will be presented. The main probiotics used, and the main benefits provided to health, will also be reported. The factors that generated the viability of these microorganisms were also reported, including pH, acidity, composition of the carrier matrix, oxygen, and temperature of the food processing stages. In addition, studies applying different probiotic microorganisms in vegetable matrices such as juices and snacks, in matrices of animal origin such as yogurt and dessert, and in matrices of animal and vegetable origin such

* Corresponding Author's Email: vanessa.c.oliveira@ufv.br.

In: Microencapsulation of Probiotics
Editors: D. de Almeida Paula, K. Cristina Guedes Silva and N. de Almeida Costa
ISBN: 979-8-88697-979-4

as bread, jelly candies, and beverages. Such studies demonstrate and reinforce the wide application of these microorganisms in the food sector.

**Keywords**: lactic bacteria, benefits, food matrix, viability, survival

## Introduction

The term probiotic is derived from the Greek word 'pro bios', which means 'life', as opposed to 'antibiotic', which means 'against life' (Guarner et al. 2005). In 1907, the Russian Nobel Prize Metchnikoff, who is considered the grandfather of modern probiotics, cited that acid-producing that the production of acid in fermented dairy products could help in the prevention of intestinal problems (Metchnikoff 1908), which is considered the first modern relatives of probiotics and the possibility of benefiting the host. The word probiotic was first used in 1953 by the German scientist Werner Kollath to name “active substances essential for a healthy development of life” (McFarland 2015). Lilly and Stillwell (1965) used the term "probiotic" in 1965, in a different context than Werner, to represent compounds produced by bacteria that promote the growth of other microbes. In 1974, Parker accepted that “probiotics are organisms and chemicals that aid intestinal balance (Parker 1974). In 1989, probiotics were tried as a "live microbial supplement that enhances host intestinal stability" by Fuller (1989).

In 2001, an Expert Consultation brought together international scientists from the FAO (Food and Nutrition Organization of the United Nations) and the WHO (World Health Organization), where probiotics were defined as “live microorganisms which, when administered in appropriate doses, confer a benefit to the health of the host” and since then, this was the term used to define probiotics. However, in 2013, the International Scientific Association of Probiotics and Prebiotics (ISAPP) held a panel of scientific and clinical experts on probiotics (with specializations in food science, gut microbiota, family medicine, microbial genetics, immunology, gastroenterology, the microbiology of probiotic bacteria and pediatrics) for a reassessment of the term used for probiotics. A grammatically correct and adequate concept for probiotics was defined as “live microorganisms which, when administered in compliance, confer a health benefit to the host” (Hill et al. 2014) which is currently used. This concept addresses the three main characteristics of probiotics, such as microbial enhancement, viability, and health (Das et al. 2022).

Several definitions of the term are interpreted with several meanings (Reid 2016). Years ago, Hasler (2002) defined probiotic bacteria as bioactive components. Probiotics have also been defined as "a microbial supplement that benefits the host, changing the microbial community associated with that host, degrading food, improving the quality of environmental characteristics, and increasing nutritional value" (Verschuere et al. 2000). Another internationally used definition of probiotics was microorganisms that confer health benefits to the host when ingested in adequate amounts (Sanders 2008), which later would be adapted correctly. According to Chua et al. (2017), probiotics are considered a "type of bacteria that colonizes our intestines and can be considered a metabolic organ due to its benefits in human health such as improvement in metabolism and immune function." The FDA (Food and Drug Administration) still uses the term for biotherapeutic probiotics aggressively. The concept of pharmacokinetics and biotherapeutics appears in the literature as a term to determine microbial activity, instead of being an alternative to the name probiotic and its definition. According to Reid (2016) the very use of the term "therapeutic" for probiotics, signals the treatment of a disease, being specific to medicine, and probiotics can be supplements, medicines, foods, medicinal foods, and others. The term pharma is understood as pharmaceutical, being like a drug or a chemical component.

In human history, the use of live microorganisms in food, mainly lactic acid bacteria, represents a history of improvement and maintenance of human health. Greeks and Romans used fermented dairy foods to keep their ancient civilizations healthy. A Roman historian in 76 BC advised the ingestion of products obtained through the fermentation of milk as a possible cure for gastroenteritis (Tegene and Kebede 2022).

The modern history of probiotics begins around 1900 through the research of Elie Metchnikoff. Elie was a scientist who worked at the Pasteur Institute in Paris, he observed that regular consumption of lactic acid bacteria from fermented milk products improved the health and longevity of a certain Bulgarian peasant population. Elie associated his discovery with the *'Bulgarian bacillus'* discovered by a Bulgarian doctor Stamen Grigov. Later, Elie claimed that the healthy bacteria in yogurt improved the immune system and aided digestion (Metchnikoff 1908; Reid, et al. 2007). At the same time as Elie, a French pediatrician, Stamen Grigorov, identified that children with diarrhea had lower amounts of bacteria with Y-shaped morphological characteristics in the composition of their feces, and in healthy children, these bacteria were found in greater quantities (Tissier 1906). Metchnikoff, stated that "the reliance of gut microbes on food makes it possible to take steps to modify the flora in our bodies and replace harmful microbes with helpful microbes." While Metchnikoff was discovering the effects of microbes on human health, Louis Pasteur discovered the microorganisms responsible for the fermentation process. These discoveries favored the development of the dairy industry in France, the first in Europe at that time, based on the use of fermented milk from *Bacillus bulgaricus* (Diplock et al. 1999; Gasbarrini, Bonvicini, and Gramenzi 2016). Yogurts and foods produced from fermented milk can be considered the first functional foods according to a consensus document created in Europe (Diplock et al. 1999).

Rettger and Cheplin (1922) performed one of the first human studies with *Lactobacillus acidophilus* in 30 patients with chronic constipation, eczema, or diarrhea. The researchers saw improvements for all the diseases studied. In another decade, Kopeloff, Blackman, and McGinn (1932) confirmed the effect of *Lactobacillus acidophilus* in individuals with mental illness and constipation. During the 1940s, research related to microorganisms focused on the identification of pathogenic bacteria, leaving aside the identification of strains of yeast or bacteria that could promote health in human life (McFarland 2015). During the 1950s to 1980s, studies on probiotics focused on identifying probiotic strains in nature or as hosts in humans and on discovering the mechanisms of action. It was possible to understand the interactions of the flora and its mechanism to resist the invasion of pathogenic bacteria being named as resistance to colonization (McFarland 2000).

## Main Microorganisms Used with Probiotic Characteristics

The most used microorganisms added to food products or for supplementation known for their probiotic characteristics include lactic acid bacteria (LAB), yeast, and bifidobacteria (Ayyash et al. 2021). The main microorganisms that have these probiotic characteristics are shown in Table 1.

*Lactobacillus* species are probably the best-known species as probiotics, but yeast as probiotics has aroused the interest of researchers. Industrial and biotechnological applications include yeast for processing fermented foods involving vitamins, enzymes, and acids. Yeasts can be characterized as unicellular eukaryotic fungi belonging to the phylum Ascomycota (such as *Saccharomyces*) and Basidiomycota (such as *Cryptococcus, and Rhodotorula*). Its reproduction can be asexual or sexual (Gil-Rodríguez, Carrascosa, and Requena 2015). Probiotic strains must present conditions to remain alive to create a symbiotic relationship in the host's gastrointestinal tract. Yeasts are very favorable to creating this relationship as they are not affected by antibacterial agents as they are resistant to antibiotics. This characteristic of yeasts makes them beneficial for labeling probiotics.

**Table 1.** Main microorganisms used as probiotics

| Microorganism | Genus | References |
|---|---|---|
| Yeasts | *Kluyveromyces lactis; Saccharomyces boulardii; Saccharomyces cerevisiae; Saccharomyces carlsbergensis; Saccharomyces pastorianus* | Holzapfel et al. (2001); Das et al. (2022); Shruthi et al. (2022); Vivek et al. (2023); Xu et al. (2022). |
| *Bifidobacterium* sp. | *Bifidobacterium longum; Bifidobacterium infantis; Bifidobacterium lactis; Bifidobacterium breve; Bifidobacterium bifidum; Bifidobacterium animals; Bifidobacterium adolascentis* | |
| *Lactobacillus* sp. | *Lactobacillus acidophilus; Lactobacillus rhamnosus; Lactobacillus gasseri; Lactobacillus casei; Lactobacillus reuteri; Lactobacillus plantarum; Lactobacillus salivarius; Lactobacillus johnsonii; Lactobacillus gallinarum; Lactobacillus plantarum; Lactobacillus fermentum; Lactobacillus helveticus; Lactobacillus brevis; Lactobacillus murinus; Lactobacillus curlatus; Lactobacillus amylovorus* | |
| Other lactic acid bacteria | *Enterococcus faecalis; Enterococcus faecium; Lactococcus lactis; Leuconostoc mesenteroides; Pediococcus acidilactici; Sporolactobacillus inulins; Streptococcus thermophilus* | |
| Secondary microorganisms | *Bacillus cereus var. toyoi; Bacillus lichniformis; Bacillus subtilis; Lactococcus lactis; Leuconostoc citreum; Pediococcus pentosaceus; Propionibacterium freudenreichii; Streptococcus infantarius; Streptococcus salivarius subsp; Escherichia colistrain nissle;* | |

About resistant *Lactobacillus*, there is the transfer of the resistant gene to the pathogenic bacteria, as gene transfer between bacteria and yeast is rare, making the application of yeast as probiotics safe (Czerucka, Piche, and Rampal 2007). Yeasts are also capable of promoting animal and human health by making minerals more bioavailable through phytate hydrolysis, detoxification of fungal and xenobiotic toxins, and folate biofortification (Afolabi, Adeyemo, and Balogun 2018).

Lactic acid bacteria (LAB) species are one of the most important species considered to be probiotics. LAB are gram-positive bacteria, that do not form spores and are considered non-aerobic, aerotolerant, highly fermentative, acid-tolerant, and lactic acid synthesizers (Axelsson and Ahrné 2000). LAB, described as probiotic microorganisms, have the characteristic of inhibiting spoilage bacteria by producing bacteriocins. These compounds are ribosomes that produce small peptides and are toxic to bacterial strains. Due to these characteristics, LAB is used in food biopreservation (HOLZAPFEL et al., 1998). These cultures are also recognized as safe, by the GRAS (Generally Recognized as Safe) by the US FDA (Wessels et al. 2004).

The vast majority of bifidobacteria species originate from the gastrointestinal tract of mammals and humans (Duranti et al. 2019). Some strains of bifidobacteria have beneficial effects on their host. This species is considered a probiotic, being health-promoting organisms, found today in large-scale functional foods and also in food supplements (Ruiz et al. 2016). The incorporation of bifidobacteria of intestinal origin in foods is limited due to their low oxygen tolerance (Simpson et al. 2005).

Microorganisms must have some general properties to be considered probiotics according to Figure 1.

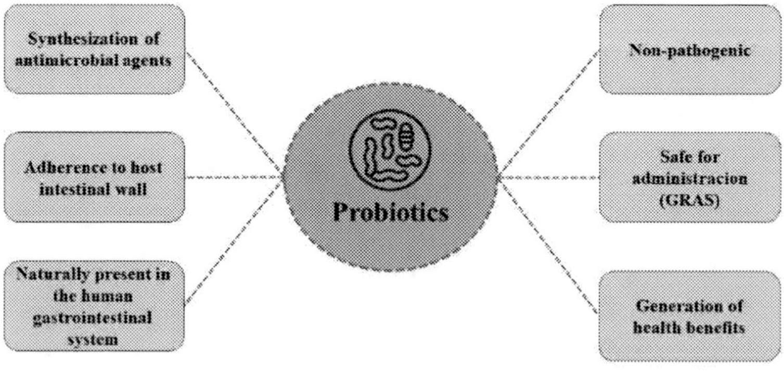


**Figure 1**. General properties of probiotic microorganisms.

## Benefits of Probiotics

The definition of health was defined by WHO (1992) as "a state of complete physical, social and mental well-being and not merely the absence of disease or infirmity." The general category of probiotics can provide essential benefits to improve the host's quality of life, promoting health. In 2013, the International Scientific Association of Probiotics and Prebiotics (ISAPP) held a debate and published a Consensus Statement where the panel agreed that certain effects can be attributed to probiotics. The statement considers that the overall benefit of probiotics is to promote healthy gut microbiota in their hosts. This overall benefit that probiotics exert on the gut microbiota is derived from the creation of a more favorable gut site, through mechanisms shared by most probiotics. Two more general benefits of probiotics were considered in this panel: the promotion of a healthy immune system and a healthy digestive tract (Hill et al. 2014). McFarland (2015) addresses that probiotic microorganisms have been tested for various indications. The researcher addressed in her literature review 420 randomized clinical trials from the years 1977 to 2014.

Lactic acid bacteria (LAB) are known to provide several host health benefits and nutritional advantages. Probiotic microorganisms can improve the absorption of nutrients from food, assist in the digestion of lactose due to the breakdown of lactase into simpler carbohydrates, regulate intestinal infections, reduce blood cholesterol and control some types of cancer (Das et al. 2022). Other species of probiotics can help in the treatment of acute and chronic infectious disorders, as well as stomach problems (Nazir et al. 2018). The body's resistance can be weakened by various factors, which can result in degenerative, viral, neoplastic, and inflammatory diseases. Antibiotic treatments, immunosuppressive therapy, and irradiation can modify the normal intestinal flora of the human body. Functional foods added with probiotics can help to accelerate the incorporation of natural cell defense mechanisms and improve cellular well-being (Zubillaga et al. 2001).

Antibiotic-associated diarrhea results from using antibiotics indiscriminately or over a long period. Antimicrobial therapy with broad-spectrum medications such as cephalosporin, clindamycin, and aminopenicillin can trigger this pathology. The use of antibiotics can disrupt the natural microbiota of the human body, because of which there is a decrease in native microorganisms. When there is a decrease in these symbiotic microorganisms, the health of the host can be unbalanced (Maziade et al. 2013; Issa and Moucari 2014; Yadav, Chandra, and Maurya 2017). Probiotics can restore intestinal microecology after antibiotic therapy. Probiotics act through receptor competition, competition for nutrients, inhibit epithelial and mucosal adherence of pathogens, induce a change in the colonic pH to a lower level, disfavoring the development of pathogenic species, and stimulate immunity and the production of antimicrobial substances (Rolf 2000). The main cause of acute childhood diarrhea is rotavirus, causing a severe gastrointestinal condition, requiring patient rehydration treatments. Along with oral rehydration treatment, probiotics may provide beneficial effects. A minimum dose of 10 billion CFU is potent to administer as therapy in children within 48 hours (Szymański et al. 2006). Approximately 20 to 60% of people traveling around the world can be affected by traveler's diarrhea. Individuals who present this pathology are those who travel from developed or industrialized areas to the most backward and remote areas (Hill and Ryan 2008). As therapy for this pathology, antibiotics present good results in prophylaxis, but cannot be recommended for generalized use, which triggers the search for options with less demand for expenses (Katelaris et al. 1995). Wilkins and Sequoia (2017) proved that probiotics can be used as a therapy for acute infectious diarrhea along with antibiotic-associated diarrhea. Another study proved that probiotics had traveler's anti-diarrheal effects (Bae, 2018).

Probiotics also have beneficial effects on the eradication rates of *Helicobacter pylori* infections (Cruz et al. 2020). Probiotics can also reduce the side effects related to the treatment of this gram-negative bacterium and reduce the gastrointestinal inflammation related to *Helicobacter pylori* (Chua et al. 2017; Gulzar et al. 2019) and can potentiate the immunoregulatory functions that work as antagonistic to the infection caused by *H. pylori* (Canducci et al. 2002; Felley and Michetti 2003; Hamilton-Miller 2003). However, the type of strain, duration, dosages, and safety of administering probiotic supplementation need further research for clinical application.

Particles that should be harmless can trigger reactions in the immune system and trigger allergies. Some strains of probiotics help in the therapy of these allergies by reducing inflammation, stabilizing the immune system, and strengthening the layer of the human intestine (Lopez-Santamarina et al. 2021). Probiotics have been used as an effective therapy for lactase deficiency in humans due to their characteristics of controlling the intestinal microbiota, assisting in the digestion of lactose, and contributing to the replacement of colonies (Das et al. 2022). These microorganisms increase the total hydrolytic capacity of the small intestine and colonic fermentation (Dhama et al. 2016).

Inflammatory bowel diseases (IBD) including ulcerative colitis (UC) and Crohn's disease (CD) chronically inflame the intestinal mucosa and are also defined as a disturbance of the intestinal barrier function and a dysbiotic microbiota, which can trigger a gastrointestinal mucosa chronic inflammation (Chibbar and Dieleman 2019). Crohn's disease and colitis are associated with the deregulation of the immune system and trigger the lack of balance of natural immunity with the environment and cultures (Matsumoto et al. 2005). The administration of probiotics can balance the immune system of the intestinal mucosa, causing a decrease in pro-inflammatory cytokines (Ishikawa et al. 2003; Kruis et al. 1997; Rembacken et al. 1999).

Recently, probiotics are effective in the therapy of metabolic diseases. In vitro and in vivo research has evaluated cholesterol reduction through the administration of probiotics (Jones et al. 2012; Martoni et al. 2015; Koppinger et al. 2020). The success of probiotics as a therapy for cholesterol-lowering and anti-inflammatory diseases must be the selection of improved strains and also the delivery technologies of the microorganisms.

Clinical studies have shown that probiotics can be effective in the treatment and prevention of viral infections (Mousa 2017; Paknahad, Moravejolahkami 2021) against common respiratory viruses such as rhinovirus, influenza, and respiratory syncytial virus (Luoto et al. 2014), then its adequate immunomodulatory activity and also the ability to stimulate the production of interferon (INF) (Bohlouli et al. 2021). It has also been suggested that administering probiotics to patients with COVID-19 may help balance the gut microbiota and reduce secondary infection through bacterial translocation (Baud et al. 2020; Xu et al. 2020). According to Galdeano et al. (2019), probiotics can also stimulate immune responses in the gut during COVID-19 infection, helping to neutralize or eliminate the virus and slow down disease progression. Probiotic microorganisms can also maintain the integrity of enterocyte junction reducing SARS-CoV2 infection, and impacting the development of COVID-19 (Baud et al. 2020). The administration of probiotics is a key therapeutic process to aid in the treatment of COVID-19, and further research is needed to determine the optimal strain(s), dosage, timing, and duration of treatment for SARS infection. -CoV-2.

Interest in probiotics is at an all-time high worldwide, driving research development, new products coming to market, and regulatory agencies to safely deliver science and technology to consumers. According to Hill et al. 2014, different parties such as regulators, industry, scientists, and consumers have specific goals in probiotic research. According to the authors, the regulatory party wants to protect consumers against misinformation, the industry wants to develop profitable and quality products containing these microorganisms, scientists want to produce quality science on the subject and consumers want safe information before consuming the products.

## Factors Affecting the Stability/Survival of Microorganisms

To be considered viable microorganisms for therapeutic use or for other purposes, such as in addition to food products, probiotics must-have attributes and characteristics necessary to be researched (Reid and Dhir 2019) such as:

1. The microorganisms must have been diagnosed alive and in sufficient quantities
2. The strains of probiotics must be genetically identified, receive numbers, letters, or names, and be named with the nomenclature in force.
3. Assignment of the strain as probiotics, as well as the strain(s) used in the host (human and animal) for the destination of the probiotic by properly schematized scientific research
4. The probability of a certain strain demonstrating efficacy in one disease and not being effective for another type of disease is high.
5. The labeling of human probiotic strains must be labeled in animal studies.

The Food and Agriculture Organization of the United Nations (FAO) and the World Health Organization (WHO) established guidelines to evaluate probiotics in food and recommended several criteria for the selection of these microorganisms, namely: the ability to adhere to the epithelium, resist the conditions unfavorable conditions that the human body can impose, present antimicrobial activity and evaluate its safety (Hill et al. 2014). Kopp-Hoolihan (2001) established other different criteria in addition to those presented above, such: as an improvement to the host and stabilization of the intestinal microflora.

The ideal functionality of a probiotic is related to its probiotic activity and its stability. In this way, probiotic strains must keep their metabolism stable and active in the food product, they must survive in adequate numbers in the digestive tract and exert beneficial effects in the host intestine, transferring the probiotic effects to the consumer (Chandramouli et al. 2004; Afzaal et al. 2022). The adequate viable cell count, that is, the "therapeutic minimum" number of probiotics for beneficial effects is from $10^6$ CFU/mL or g to $10^8$ CFU/mL or g per day (Minelli and Benini 2008).

The survival of probiotic microorganisms is dependent on several factors that may occur during the processing and formulation of the food product, such: as redox environment, pH, presence of other microorganisms that may impair the survival of probiotics, temperature, and storage time, the water activity of the product, the concentration of acetic and lactic acids, hydrogen peroxide, titratable acidity (Mattila-Sandholm et al. 2002; Vivek et al. 2023). Farnworth and Champagne (2016) consider other factors that must be addressed before the incorporation of probiotic microorganisms into foods that could compromise viability and survival, such: as verifying the safety of the food about cross-contamination, the compatibility of the probiotic with the food product, verification of the viability and stability of the probiotic strain during processing, packaging, and storage, verification of the pH of the product about the survival of the probiotic strain and physicochemical fermentation conditions if this is objective about the final product.

Oxygen is a powerful oxidizing compound that can affect the viability/survival of probiotics. The Eh value (the ability of a molecule to reduce or oxidize another molecule) has a drastic effect on the viability of probiotic microorganisms during the processing and storage of a food product (Storz and Imlay 1999; Talwalkar and Kailasapathy 2004). For example, dairy-based food products, can freeze, reduce oxygen toxicity and promote probiotic constancy (Fenster et al. 2019).

The acidity present in the food product can alter the viability of probiotic microorganisms. Inactivation of probiotics can occur in acidic conditions due to damage to cell membranes that are caused by high levels of hydrogen ions present in the food matrix (Lo Curto et al. 2011).

The composition of the probiotic carrier food matrix can also influence the viability of the strains, which may or may not serve as an energy source for their development and maintenance. The physical state of the food matrix also influences the survival of probiotics. Foods in liquid states are digested faster than foods in solid states, consequently decreasing the contact time of probiotics with stressors, such as stomach acid (Huang and Adams 2004; Klu and Chen 2015). Thus, the development of probiotic foods, especially those with acidic pH characteristics, low water activity or that may be exposed to heat treatments with temperatures above 60°C, can be a great challenge for the industry. Lactic acid bacteria (LAB), for example, cannot survive for a long time under these proposed conditions (Cutting 2011).

Other factors also influence the viability of probiotic microorganisms in products during pre-processing, processing, and storage steps. Factors observed include processing parameters (heat treatment, incubation temperature, product cooling rate, packaging materials, storage methods, and scale of production); and microbiological parameters (strains of probiotics, rate, and proportion of inoculation) (Tripathi and Giri 2014). Adaptations may be necessary for the steps of adding probiotics to foods and in processing techniques. As an alternative, probiotic strains may be included after the thermal processes performed on the food or alternatives to protect these cells from probiotic microorganisms (El Kadri et al. 2018; Favaro-Trindade and De Pinho 2008; Gutiérrez-Zamorano et al. 2019).

Probiotic microencapsulation is a technique that emerges as an additional technology to protect probiotic microorganisms from losing their viability and maintain their survival during processing, storage, and gastrointestinal transit (Xu et al. 2022). It is defined as a process where tiny droplets of active substances are coated with a tiny capsule. This capsule is commonly referred to as a shell capsule, external phase, membrane, matrix, or, carrier material (Nedovic et al. 2011). Microencapsulation protects active materials against various physicochemical stresses, such as humidity, heat, pH, and harmful substances (Islam et al. 2010; Solanki et al. 2013).

## Main Food Matrixes with the Addition of Probiotics

The production of foods with probiotic claims represents a challenge for the industry, as there are still some difficulties in maintaining the survival of the probiotic strains added to foods during processing, storage, distribution, and until the food reaches the consumer (Min et al. 2019).

Historically, the best carriers of probiotic microorganisms are fermented milk products. These products are so named because they excel in being a source of nutrients to carry or generate live and active cultures. Over time, optimization occurred to promote cell survival in the fermentation process. Dairy foods, especially fermented dairy products such as yogurt, are effortlessly adapted to maintain the viability of the microorganisms of interest (Heller 2001; Kandylis et al. 2016). Other dairy products such as cheese and acidified milk are also accepted as carrier vehicles for probiotic delivery. The physicochemical and nutritional characteristics of these products provide the buffering capacity in adverse conditions such as the stomach with a pH of 2-3, for a considerable number of probiotic survival in the lower intestine (6 to

8 log CFU/mL) that can benefit the cells to exert therapeutic effects on the host (Araújo et al. 2012; Kalicka et al. 2019; Linares et al. 2017; Reid 2015; Tomar 2019). But an expressive amount of the population presents, does not consume these products, due to restrictions of dietary factors, due o allergies and intolerances to dairy products, and the inability to consume this food niche. There is also an increase in people who have adopted vegetarian and vegan lifestyles, among others. This new demand led industry and research to develop non-dairy probiotic foods of plant origin such as fruits, vegetables, and cereals (Pinto and Vilela 2021). Non-dairy food products have unique characteristics and advantages such as avoiding allergens present in dairy products; providing low-cholesterol products to consumers suffering from cardiovascular disease or obesity; adequacy of eating habits of different ethnic groups; improving the flavor of soy products; increasing the nutritional value of non-dairy foods (Granato et al. 2010; Kołożyn-Krajewska and Dolatowski 2012; Nagpal et al. 2012; Stadnik and Dolatowski 2014; Vasudha and Mishra 2013; Yeo and Liong 2010). The functional properties of plant matrices and their structure make them suitable for transporting probiotic strains that would benefit from the characteristics of plant matrices. The functional characteristics depend on its richness in nutrients, fibers, vitamins, minerals, and bioactive compounds. Some of these components perform interactions with intestinal microorganisms (Flach et al. 2018). Plant tissues are composed of various internal microstructures of cells, pores, and intercellular spaces, where microorganisms can adhere and act as carriers of probiotic cells (Peres et al. 2012), being considered ideal structures to maintain the survival of probiotic strains. In Table 2 and Table 3, different products with different food matrices were grouped, for the incorporation of probiotic microorganisms.

The available data indicate that the intrinsic properties of different food matrices or a combination of them can be increased successfully explored and develop a partnership with probiotic microorganisms. The benefits and viability of probiotics in food products were improved and modulated by the food matrix which was inserted, in most cases a final product with functional characteristics was obtained, having a synergistic effect of the probiotic strain and the matrix. For the selection of the food matrix, several aspects must be considered to obtain a suitable final product for a given probiotic strain, which is capable of providing all the benefits of probiotic microorganisms (De Bellis, Sisto, and Lavermicocca 2021).

**Table 2.** Main vegetable food matrices associated with probiotic bacteria

| Food matrix | Probiotic added | Main results | Reference |
|---|---|---|---|
| Freeze-dried banana powder (rehydrated drink) | *Lacticaseibacillus paracasei* | Probiotic freeze-drying preserved the high probiotic cell survival capacity (1011 CFU/g). On storage for 60 days, the powdered product was stable and maintained probiotic survival (107 CFU/g) | Sornsenee et al. (2022) |
| Table olive paste | *Lactobacillus plantarum* 33 | Encapsulation conferred additional protection to L. plantarum 33 (about 7 log CFU/g) when exposed to simulated gastrointestinal conditions | Alves et al. (2015) |
| Yacon root | *Lactobacillus casei* LC-01 | Yacon dry root was evaluated as a potential probiotic carrier. Bacteria adhere to the surface of dried yacon to survive gastric conditions. | Leone et al. (2017) |
| Purple pitaya juice (*Hylocereus* sp.) | *Lactobacillus rhamnosus* *Lactobacillus paracasei* | *Lactobacillus rhamnosus* survives longer than *Lactobacillus paracasei* in pitaya juice. | Usaga et al. (2022) |
| Mango Snacks | *Lactiplantibacillus plantarum* CIDCA 83114 | In in vitro digestion, probiotic counts were > 7 log CFU/g snack in the salivary, gastric, and intestinal phases. Bacterial counts throughout storage were > 6log CFU/g snack under all conditions tested. | La Cava et al. (2022) |
| Beverage based on barley, ragi, moth beans, soy, almond, and coconut | *Lactobacilli acidophilus* | Sprouted drink mix probiotic drinks have higher probiotic counts than non-sprouted drink mix probiotic drinks. | Chavan et al. (2018) |
| Kiwi juice | *Lactobacillus acidophilus 85* *Lactobacillus helveticus 76* *Lactobacillus plantarum 90* | Kiwi juices were favorable matrices for LAB growth and colony counts remained above 9.0 log CFU/mL after fermentation. | Wang et al. (2022) |

**Table 3.** Main animal food matrices associated with probiotic bacteria

| Food matrix | Probiotic added | Main results | Reference |
|---|---|---|---|
| Functional probiotic yogurt with royal jelly fortification | *Streptococcus thermophilus Lactobacillus bulgaricus* | Yogurt added with royal jelly was more acceptable than samples without the jelly in rheological properties, the viability of probiotic cultures, and, fatty acid composition. | Kavas (2022) |
| Petit suisse cheese | *Lactobacillus acidophilus*<br><br>*Bifidobacterium* | The commercial sample showed higher scores in all aspects by consumers, while the probiotic petit suisse samples showed the opposite behavior. | Pereira et al. (2016) |
| Creamy milk chocolate dessert | *Lactobacillus paracasei* | The reduction in pH and the increase in acidity did not interfere with the viability of *L. paracasei*, which remained in amounts greater than 8 log CFU/g during the 28-day storage period | Valencia et al. (2016) |
| Cheesecake | *Lactobacillus rhamnosus* HN001 | *L. rhamnosus* HN001 can survive the storage period at levels greater than 6 log CFU/g. | Hasgucmen and Sengun (2020) |
| Dairy drinks (buffalo milk and cow's milk) | *Streptococcus thermophilus Lactobacillus bulgaricus Lactobacillus acidophilus* | Lactic acid bacteria showed viable cell counts at the end of shelf life, with the highest values (7.33 to 8.83 log CFU/mL) detected in buffalo dairy products. | Silva et al. (2020) |
| Fat-free natural yogurt | *Lactobacillus acidophilus* | Yogurts containing prebiotics of different chain lengths had L. acidophilus counts comparable to each other, but higher counts than the control. | Aryana et al. (2007) |
| Goat ricotta | *Lactobacillus acidophilus Bifidobacterium lactis* | In the in vitro test, counts of each of the probiotic strains tested were approximately 6 log CFU/g in goat ricotta. | Meira et al. (2015) |
| Plant-based fermented milk drink with soy, almond, and milk | *Lactobacillus rhamnosus Lactobacillus acidophilus Lactobacillus plantarum Lactobacillus casei* | Fermented soy and almond milk and their combination may serve as an effective vehicle for probiotics with antioxidant activity. | Zahrani and Shori (2023) |

**Table 3.** (Continued)

| Food matrix | Probiotic added | Main results | Reference |
|---|---|---|---|
| Probiotic gelatin candies enriched with juçara pulp and passion fruit | *Bacillus coagulans* GBI-30 6086 | *B. coagulans* maintained counts greater than 6.4 Log CFU/g for 90 days. In vitro, there were 6.82 Log CFU/g in enteral phase II. Consumption of a 30 g portion of the candies yielded approximately 7.82 Log CFU/g of the probiotic. | Miranda et al. (2020) |
| Low-fat yogurt added with vegetable extract (olives, garlic, onions, and citrus fruits with sodium acetate as a vehicle) | *Lactobacillus bulgaricus* *Streptococcus thermophilus* | The 0.5% plant extract increased the viability of *L. bulgaricus* in skimmed yogurt, without affecting the physicochemical characteristics. | Michael, Phebus, and Schmidt (2010) |
| Whipped yogurt enriched with flaxseed | *Lactobacillus acidophilus* | The addition of flaxseed to the probiotic yogurt caused an increase in the growth of L. acidophilus (up to 8.82 CFU/mL) compared to the control sample (6.87 CFU/mL). | Mousavi et al. (2019) |
| Whipped yogurt enriched with aloe vera leaf gel | *Lactobacillus Acidophilus* *Bifidobacterium lactis Bb-12* | Aloe vera leaf gel had a negative effect on microorganisms, but viable cells were in the acceptable range of ($10^6$ –$10^7$ CFU g $^{-1}$). | Azari-Anpar et al. (2017) |
| Functional bread | *Lactobacillus acidophilus* | *L. acidophilus* included in the microcapsules can be incorporated into the bread surface through edible coatings, resulting in functional bread with characteristics similar to regular bread, but with additional health benefits. | Altamirano -Fortoul et al. (2020) |
| Whey drink with acai pulp | *Bifidobacterium longum* BI-05 *Lactobacillus acidophilus* La-14 | After 21 days of storage, the drink showed higher counts ($10^7$ and $10^8$ CFU/mL) for both microorganisms. | Zoellner et al. (2009) |

## Conclusion

The research and the market for probiotics continue to grow, either by increasing the variety of products or by the increase in people who use them according to the benefits that are proven every day. But the challenges for researchers, the food industry, health professionals, and the public are great regarding product regulation and administration guidelines for consumers.

## References


Afolabi, Folake Titilayo, Adeyemo, Stella Mojisola, and Balogun, Halimat Osheiza. 2018. "Fermentation Conditions and Process Optimization of Citric Acid Production by Yeasts." *The International Journal of Biotechnology* 7:51-63.

Afzaal, Muhammad, Saeed, Farhan, Ahmed, Aftab, Saeed, Muhammad, and Ateeq, Huda. 2022. "Hydrogels as the carrier for the delivery of probiotics" In J. Singh, A. Vyas (Eds.), *Advances in Dairy Microbial Products*, Woodhead Publishing. 303-315. https://doi.org/10.1016/B978-0-323-85793-2.00019-9

Ahmad, Asif and Khalid Sumaira. 2018. "Therapeutic aspects of probiotics and prebiotics." *Diet, Microbiome and Health*. 53-91. https://doi.org/10.1016/B978-0-12-811440-7.00003-X

Altamirano-FortouL, R., Moreno-Terrazas, Rubén, Quezada-Gallo, A. and Rossell, Cristina M. 2012. "Viability of some probiotic coatings in bread and its effect on the crust mechanical properties." *Food Hydrocolloids*, 29:166-174. https://doi.org/10.1016/j.foodhyd.2012.02.015

Alves, Marta, Peres, Cátia M., Hernandez-Mendonza, Adrián, Bronze, M. Rosário, Peres, Cidália and Malcata, F. Xavier. 2015. "Olive paste as a vehicle for delivery of potential probiotic *Lactobacillus plantarum* 33." *Food Research International*, 75:61-70. https://doi.org/10.1016/j.foodres.2015.04.048

Anukan, Kingsley C and Gregor Reid. 2007. "Probiotics: 100 years (1907-2007) after Elie Metchnikoff's." *Communicating Current Research and Educational Topics and Trends in Applied Microbiology*. Edit A. Méndez-Vilas, 467-74.

Araújo, Emiliane Andrade, Pires, Ana Clarissa dos Santos, Pinto, Maximiliano Soares, Jan Gwénael and Carvalho, Antônio Fernandes de. 2012. "Probiotics in dairy fermented products." In: Rigobelo, E. C. (Ed.), *Probiotics, Rennes*: CC." United Kingdom, 129-148.

Aryana, K. J, Plauche, S., Rao, R. M., McGrew. P. and Shah, N. P. 2007. "Fat-Free Plain Yogurt Manufactured with Inulins of Various Chain Lengths and *Lactobacillus acidophilus." Journal of Food Science*, 72:79-84. https://doi.org/10.1111/j.1750-3841.2007.00302.x

Axelsson, Lars and AHRNÉ, Siv. 2000. "Lactic Acid Bacteria." Priest, F. G., Goodfellow, M.; (Eds.), *Applied Microbial Systematics*, Springer, Dordrecht, p. 367-88.

Ayyash, Mutamed M., Abdalla, Abdelmoneim K., Alkalbani, Nadia S., Baig, Mohd Affan, Turner, Mark S., Liu, Shao-Quan and Shah, Nagendra P. 2021. "Invited review: Characterization of new probiotics from dairy and nondairy products—Insights into acid tolerance, bile metabolism and tolerance, and adhesion capability." *Journal of Dairy Science*, 104;8363-379. https://doi.org/10.3168/jds.2021-20398

Azari-Anpar, Mojtaba, Payeinmahali, Hadiseh, Garmakhany, Amir Daraei and Mahounak, Alireza Sadeghi. 2017. "Physicochemical, microbial, antioxidant, and sensory properties of probiotic stirred yoghurt enriched with *Aloe vera* foliar gel." *Journal of Food Processing and Preservation*, 41:e13209. https://doi.org/10.1111/jfpp.13209

Bae, Jong-Myon. 2018. "Prophylactic efficacy of probiotics on travelers' diarrhea: An adaptive meta-analysis of randomized controlled trials." *Epidemiology and Health*, 40:e2018043. https://doi.org/10.4178/epih.e2018043

Baud, David, Dimopoulou, Varvara Agri, Gibson, Glenn R., Reid, Gregor, and Giannoni, Eric. 2020. "Using probiotics to flatten the curve of coronavirus disease COVID-2019 pandemic." *Frontiers in Public Health*, 8:186. https://doi.org/10.3389/fpubh.2020.00186

Bohlouli, Jalal, Namjoo, Iman, Borzoo-Isfahani, Mohammad, Kermani, Mohammad Ali Hojjati, Zehi, Zakiyeh Balouch and Moravejolahkami, Amir Reza. 2021."Effect of probiotics on oxidative stress and inflammatory status in diabetic nephropathy: a systematic review and meta-analysis of clinical trials." *Heliyon*, 7:e05925. https://doi.org/10.1016/j.heliyon.2021.e05925

Canducci, F., Cremonini, F.; Armuzzi, A., Di Caro, S., Gabrielli, M., Santarelli, L., Nista, E, Lupascu, A., De Martini, D. and Gasbarrini, A. 2002. "Probiotics and *Helicobacter pylori* eradication." *Digestive and Liver Disease*, 34:81-83. https://doi.org/10.1016/S1590-8658(02)80172-4

Chandramouli, V., Kailasapathy, K., Peiris, P. and Jones, M. 2004. "An improved method of microencapsulation and its evaluation to protect *Lactobacillus* spp. in simulated gastric conditions." *Journal of Microbiological Methods*, 56:27-35. https://doi.org/10.1016/j.mimet.2003.09.002

Chavan, Mayuri, Gatb, Yogesh, Harmalkar, Mugdha and Waghmare, Roji. 2018. "Development of non-dairy fermented probiotic drink based on germinated and ungerminated cereals and legume." *LWT - Food Science and Technology*, 91: 339-344. https://doi.org/10.1016/j.lwt.2018.01.070

Chibbar, Richa Dieleman, Levinus A. 2019. "The gut microbiota in celiac disease and probiotics." *Nutrients*, 11:2375. https://doi.org/10.3390/nu11102375

Chua, Koon Jiew, Kwok, Wee Chiew, Aggarwal, Nikhil, Sun, Tao and Chang, Mattew Wook. 2017. "Designer probiotics for the prevention and treatment of human diseases." *Current Opinion in Chemical Biology*, 40:8-16. https://doi.org/10.1016/j.cbpa.2017.04.011

Cruz, Bruna C. S, Sarandy, Mariáurea M., Messias, Anny C., Gonçalves, Reggiani V., Ferreira, Célia L. L. and Peluzio, Maria C. G. 2020. "Preclinical and clinical relevance of probiotics and symbiotics in colorectal carcinogenesis: A systematic review." *Nutrition Reviews*, 78:667-87. https://doi.org/10.1093/nutrit/nuz087

Cutting, Simon M. 2011. "*Bacillus* probiotics." *Food Microbiology*, 28:214-220. https://doi.org/10.1016/j.fm.2010.03.007

Czerucka, D., Piche, T., Rampal, P. 2007. "Review article: yeast as probiotics - *Saccharomyces boulardii*." *Alimentary Pharmacology & Therapeutics*, 26:767-778. https://doi.org/10.1111/j.1365-2036.2007.03442.x

Das, Tridip K., Pradhan, Shrabani, Chakrabarti, Sudipta, Mondal, Keshab Chandra, and Ghosh, Kuntal. 2022. "Current status of probiotics and related health benefits." *Applied Food Research*, 2:100185. https://doi.org/10.1016/j.afres.2022.100185

De Bellis, Palmira, Sisto, Angelo and Lavermicocca, Paola.2021. "Probiotic bacteria and plant-based matrices: An association with improved health-promoting features." Journal of Functional Foods, 87:104821. https://doi.org/10.1016/j.jff.2021.104821

Dhama, Kuldeep, Latheef, Shyma K., Munjal, Ashok K., Khandia, Rekha, Samad, Hari A., Iqbal, Hafiz M. N. and Joshi, Sunil K. 2016. "Probiotics in curing allergic and inflammatory conditions-research progress and futuristic vision." *Recent Patents on Inflammation & Allergy Drug Discovery*, 10:105-118.

Diplock, A. T., Aggett, P. J., Ashwell M., Bornet, F., Fern, E. and Roberfroid, M. 1999. "Scientific concepts of functional foods in Europe consensus document." *British Journal of Nutrition*, 81:1-27.

Duranti, Sabrina, Lugli, Gabriele A., Napoli, Stefania, Anzalone, Rosaria, Milani, Christian, Mancabelli, Leonardo, Alessandri, Giulia, Turroni, Francesca, Ossiprandi, Maria Cristina; van Sinderen, D. and Ventura, Marco. 2019. "Characterization of the phylogenetic diversity of five novel species belonging to the genus *Bifidobacterium*: *Bifidobacterium castor* sp. nov., *Bifidobacterium callimiconis* sp. nov., *Bifidobacterium goeldii* sp. nov., *Bifidobacterium samirii* sp. nov. and B." *International Journal of Systematic and Evolutionary Microbiology*, *69*:1288–298. https://doi.org/10.1099/ijsem.0.003306

El Kadri, Hani, Lalou, Sofia, Mantzouridou, Fanith and Gkatzionis, Konstantinos. 2018. "Utilisation of water-in-oil-water (W1/O/W2) double emulsion in a set-type yogurt model for the delivery of probiotic *Lactobacillus* paracasei." *Food Research International*, 107:325-336. https://doi.org/10.1016/j.foodres.2018.02.049

Farnworth, Edward, Champagne, P. C. 2016. "Production of Probiotic Cultures and Their Incorporation into Foods." In: *Probiotics, Prebiotics, and Synbiotics: Bioactive Foods in Health Promotion*. 303-18. https://doi.org/10.1016/B978-0-12-374938-3.00001-3

Favaro-Trindade, Carmen, Pinho, Samantha and Rocha, Glaucia Aguiar. 2008. "Revisão: Microencapsulação de ingredientes alimentícios." - Review: Microencapsulation of food ingredientes. *Brazilian Journal of Food Technology*, 11:103-112.

Felley, C., and Michetti. P.2003."Probiotics and *Helicobacter pylori*." *Best Practice & Research: Clinical Gastroenterology*, 17:785-791. https://doi.org/10.1016/S1521-6918(03)00070-2

Fenster, Kurt, Freeburg, Barbara, Hollard, Chris, Wong, Connie, Laursen, Rune Ronhave and Ouwehand Arthur C. 2019. "The production and delivery of probiotics: A review of a practical approach." *Microorganisms* 7:83. https://doi.org/10.3390/microorganisms7030083

Flach, Joost, Van Der Waal, Mark B, Van Den Nieuwboer, Maurits, Claassen, Eric and Larsen, Olaf F. A. 2018. "The underexposed role of food matrices in probiotic products: Reviewing the relationship between carrier matrices and product parameters." *Critical Reviews in Food Science and Nutrition*, 58:2570-584. https://doi.org/10.1080/10408398.2017.133462

Food & Organization, A.O.W.H. 2001. "Report of a joint FAO/WHO expert consultation on the evaluation of health and nutritional properties of probiotics in food including powder milk with live lactic acid bacteria." Argentina: *FAO/WHO.*

Fuller, R. 1989. "A review: Probiotics in man and animals." *Journal of Applied Bacteriology*, 66:365-378.

Galdeano, Carolina Maldonado, Cazorla, Silvia Inés, Dumit, José María Lemme, Vélez, Eva and Perdigón, Gabriela. 2019. "Beneficial effects of probiotic consumption on the immune system." *Annals of Nutrition and Metabolism*, 74:115-124. https://doi.org/10.1159/000496426

Gasbarrini, Giovanni, Bonvicini, Fiorenza, Gramenzi, A. 2016. "Probiotics History." *Journal of Clinical Gastroenterology*, 50:116-119. https://doi.org/10.1097/MCG.0000000000000697

Gil-Rodríguez, Ana María, Carrascosa, Alfonso and Requena, Teresa. 2015. "Yeasts in foods and beverages: in vitro characterization of probiotic traits." *LWT - Food Science and Technology*, 64:1156-162. https://doi.org/10.1016/j.lwt.2015.07.042

Granato, Daniel, Branco, Gabriel F., Nazzaro, F, Cruz, Adriano G. and Faria, José A. F. 2010. "Functional foods and nondairy probiotic food development: Trends, concepts, and products." *Comprehensive Reviews in Food Science and Food Safety*, 9:292–302. https://doi.org/10.1111/j.1541-4337.2010.00110.x

Guarner, F., Perdigon, G., Corthier, G., Salminen, S., Koletzko, B. and Morelli L. 2005. "Should yoghurt cultures be considered probiotic?" *British Journal of Nutrition*, 93:783-86. https://doi.org/10.1079/BJN20051428

Gutiérrez-Zamorano, Cristian, González-Ávila, Marisela, Díaz-Blas, Gustavo, Smith, Carlos. T., González-Correa, Carlos and García-Cancino, Apolinaria. 2019. "Increased anti-helicobacter pylori effect of the probiotic *Lactobacillus fermentum* UCO-979C strain encapsulated in carrageenan evaluated in gastric simulations under fasting conditions." *Food Research International*, 121:812-16. https://doi.org/10.1016/j.foodres.2018.12.064

Hamilton-Miller, J. M. 2003. "The role of probiotics in the treatment and prevention of *Helicobacter pylori* infection." *International Journal of Antimicrobial Agents*, 22:360-366. https://doi.org/10.1016/S0924-8579(03)00153-5

Hasgucmen, Ceren Kutlu, Sengun, Ilkin Yucel. 2020. "Viability of probiotic strain *Lactobacillus rhamnosus* and its impact on sensory properties of cheesecake during storage at −20 C and 4 C." *LWT - Food Science and Technology*, 134: 109967. https://doi.org/10.1016/j.lwt.2020.109967

Hasler, Clare M. 2002. "Functional foods: benefits, concerns, and challenges—a position paper from the American Council on Science and Health." *The Journal of Nutrition*, 132:3772-3781. https://doi.org/10.1093/jn/132.12.3772

Heller, K. J. 2001. "Probiotic bacteria in fermented foods: product characteristics and starter organisms." *The American Journal of Clinical Nutrition*, 73:374-379. https://doi.org/10.1093/ajcn/73.2.374s

Hill, Colin, Guarner, Francisco, Reid, Gregor, Gibson, Glenn R., Merenstein, Daniel J, Pot, Bruno, Morelli, Lorenzo, Canani, Roberto Berni, Flint, Harry J., Salminen, Seppo, Calder, Philip C, and Sanders Mary Ellen. 2014. "Expert consensus document: The International Scientific Association for Probiotics and Prebiotics consensus statement on the scope and appropriate use of the term probiotic." *Nature Reviews Gastroenterology & Hepatology,* 11:506-514. https://doi.org/10.1038/nrgastro.2014.66

Hill, David R. and Ryan, Edward T. 2008. "Management of travelers' diarrhoea." *British Medical Journal*, 337:a1746. https://doi.org/10.1136/bmj.a1746

Holzapfel, Wilhelm H, Haberer, Petra, Geisen, Rolf, Björkroth, Johanna and Schillinger, Ulrich. 2001. "Taxonomy and important features of probiotic microorganisms in food and nutrition." *American Journal of Clinical Nutrition*, 73:365-373. https://doi.org/10.1093/ajcn/73.2.365s

Holzapfel, W. H., Haberer, P., Snel, J., Schillinger, U. and Veld, J. H. H. i'nt. 1998. "Overview of gut flora and probiotics." *International journal of food Microbiology*, 41:85-101. https://doi.org/10.1016/S0168-1605(98)00044-0

Huang, Yang and Adams, Michelle C. 2004. "*In vitro* assessment of the upper gastrointestinal tolerance of potential probiotic dairy propionibacteria." *International Journal of Food Microbiology*, 91:253-260. https://doi.org/10.1016/j.ijfoodmicro.2003.07.001

Ishikawa, Hideki; Akedo, Ikuko, Umesaki, Yoshinori, Tanaka, Ryuichiro, Imaoka, Akemi and Otani, Toru. 2003. "Randomized controlled trial of the effect of bifidobacteria-fermented milk on ulcerative colitis." *Journal of the American College of Nutrition*, 22:56-63. https://doi.org/10.1080/07315724.2003.10719276

Islam, Mohammad Ariful, Yun, Cheol-Heui, Choi, Yun-Jaie, and Cho, Chong-Su. 2010. "Microencapsulation of live probiotic bacteria." *Journal of Microbiology and Biotechnology,* 20:1367-377.

Issa, Iyad and Moucari, Rami. 2014. "Probiotics for antibiotic-associated diarrhea: do we have a verdict?" *World Journal of Gastroenterology*, 20:17788-795. https://doi.org/10.3748/wjg.v20.i47.17788

Jones, Mitchell L, Martoni, Christopher J., Parent, Mathieu, and Prakash, Satya. 2012."Cholesterol-lowering efficacy of a microencapsulated bile salt hydrolase-active *Lactobacillus reuteri* NCIMB 30242 yoghurt formulation in hypercholesterolaemic adults." *The British Journal of Nutrition*, 107:1505–1513. https://doi.org/10.1017/S0007114511004703

Kalicka, Dorota, Znamirowska, Agata, Pawlos, Malgorzata, Buniowska, Magdalena, and Szajnar, Katarzyna. 2019."Physical and sensory characteristics and probiotic survival in ice cream sweetened with various polyols." *International Journal Dairy Technology*, 72:456-465. https://doi.org/10.1111/1471-0307.12605

Kandylis, Panagiotis, Pissaridi, Katerina, Bekatorou, Argyro, Kanellaki, Maria, Koutinas, Athanasios A. 2016. "Dairy and non-dairy probiotic beverages." *Current Opinion in Food Science*, 7:58-63. https://doi.org/10.1016/j.cofs.2015.11.012

Katelaris, P. H., Salam, I. and Farthing, M. J. 1995. "Lactobacilli to prevent traveler's diarrhea?" *New England Journal of Medicine*, 333:1360-61. https://doi.org/10.1056/NEJM199511163332016

Kavas, Nazan. 2022. "Functional probiotic yoghurt production with royal jelly fortification and determination of some properties." *International Journal of Gastronomy and Food Science*, 28:100519. https://doi.org/10.1016/j.ijgfs.2022.100519

Klu, Yaa Asantewaa Kafui and Chen, Jinru. 2015. "Effect of peanut butter matrices on the fate of probiotics during the simulated gastrointestinal passage." *LWT - Food Science and Technology*, 62:983-988. https://doi.org/10.1016/j.lwt.2015.02.018

Kołożyn-Krajewska and Danuta, Dolatowski, Zbigniew J. 2012. Probiotic meat products and human nutrition." *Process Biochemistry*, 47:1761–72. https://doi.org/10.1016/j.procbio.2012.09.017

Kopeloff, N., Blackman, N., McGinn, B. 1932. "The Incidence of *Lactobacillus acidophilus* in adults." The Journal of Infectious Diseases, 50:426-29.

Kopp-Hoolihan, Lori. 2001. "Prophylactic and therapeutic uses of probiotics: a review." *Journal of the American Dietetic Association*, 101:229-41. https://doi.org/10.1016/S0002-8223(01)00060-8

*Koppinger, Matthew Perry, Lopez-Pier, Marissa Anne, Skaria, Rinku, Harris, Preston Royal and Konhilas, John P. 2020. "Lactobacillus reuteri* attenuates cardiac injury without lowering cholesterol in low-density lipoprotein receptor-deficient mice fed standard chow." *American Journal of Physiology-Heart Circulatory Physiology*, 319: 32-41. https://doi.org/10.1152/ajpheart.00569.2019

Kruis, W., Schutz, E., Fric, P., Fixa, B., Judmaier, G. and Stolte, M. 1997. "Double-blind comparison of an oral *Escherichia coli* preparation and mesalazine in maintaining remission of ulcerative colitis." *Alimentary Pharmacology & Therapeutics*, 11: 853-858. https://doi.org/10.1046/j.1365-2036.1997.00225.x

La Cava, Enzo, Gerbino, Esteban, Sgroppo, Sonia and Gomez-Zavaglia, A. 2022. "Baked mango slices coated with *Lactiplantibacillus plantarum* immobilized in pectic extracts." *Applied Food Research*, 2:100236. https://doi.org/10.1016/j.afres.2022.100236

Leone, Roberta de Souza, Andrade, Eriel Forville de, Ellendersen, Luciana Neves, Cunha, Amanda Tais da, Martins, Ana Maria Chupel Martins, Granato, Daniel and Masson, Maria Lúcia. 2017. "Evaluation of dried yacon (*Smallanthus sonchifolius*) as an efficient probiotic carrier of *Lactobacillus casei* LC-01."*LWT - Food Science and Technology*, 75: 220-26. https://doi.org/10.1016/j.lwt.2016.08.027

Lilly, Daniel M. and Stillwell, Rosalie H. 1965. "Probiotics: Growth-promoting factors produced by microorganisms." *Science*, 147:747-48. https://doi.org/10.1126/science.147.3659.747

Linares, Daniel M., Gomez, Carolina, Renes, Erica, Fresno, José. M., Tornadijo, María E., Ross, R. P. and Stanton, Catherine. 2017. "Lactic acid bacteria and bifidobacteria with potential to design natural biofunctional health-promoting dairy foods." *Frontiers in Microbiology*, 8:846. https://doi.org/10.3389/fmicb.2017.00846

Lo Curto, Alberto, Pitino, Iole, Mandalari, Giuseppina, Dainty, Jack Richard, Faulks, Richard Martin and John Wickham, Martin Sean. 2011. "Survival of probiotic lactobacilli in the upper gastrointestinal tract using an in vitro gastric model of digestion." *Food Microbiology*, 28:1359-366. https://doi.org/10.1016/j.fm.2011.06.007

Lopez- Santamarina, Aroa, Gonzalez, Esther Gonzalez, Lamas, A., Mondragon, Alicia Del Carmen, Regal, Patricia and Miranda, Jose Manual. 2021. "Probiotics as a possible strategy for the prevention and treatment of allergies. A narrative review." *Foods*, 10:701. https://doi.org/10.3390/foods10040701

Luoto, Raakel. Ruuskanen, Olli, Waris, Matti, Kalliomäki, Marko, Salminen, Seppo and Isolauri, Erika. 2014. "Prebiotic and probiotic supplementation prevents rhinovirus infections in preterm infants: a randomized, placebo-controlled trial." *The Journal of Allergy and Clinical Immunology*,133:405-13. https://doi.org/10.1016/j.jaci.2013.08.020

Martoni, Christopher J., Labbé, Alain, Ganopolsky, Jorge G., Prakash, Satya and Jones, Mitchell L. 2015. "Changes in bile acids, FGF-19, and sterol absorption in response to bile salt hydrolase active *L. reuteri* NCIMB 30242." *Gut Microbes*, 6:57–65. https://doi.org/10.1080/19490976.2015.1005474

Matsumoto, Satoshi, Hara, T., Hori, T., Mitsuyama, K., Nagaoka, M., Tomiyasu, N., Suzuki, A., Sata, M. 2005. "Probiotic Lactobacillus-induced improvement in murine chronic inflammatory bowel disease is associated with the down-regulation of pro-inflammatory cytokines in lamina propria mononuclear cells." *Clinical and Experimental Immunology*, 140:417-26. https://doi.org/10.1111/j.1365-2249.2005.02790.x

Mattila-Sandholm, Tiina, Myllärinen, Paivi, Crittenden, Ross, Mogensen, G., Fondén, Rangne and Saarela, M. H. 2002. "Technological challenges for future probiotic foods." *International Dairy Journal,* 12:173-182. https://doi.org/10.1016/S0958-6946(01)00099-1

Maziade, Pierre-Jean, Andriessen, Anneke, Pereira, P., Currie, Brian and Goldstein, Ellie J. C. 2013. "Impact of adding prophylactic probiotics to a bundle of standard preventative measures for *Clostridium difficile* infections: enhanced and sustained decrease in the incidence and severity of infection at a community hospital." *Current Medical Research and Opinion*, 29:1341-47. https://doi.org/10.1185/03007995.2013.833501

McFarland, Lynne. 2000. "Normal flora: diversity and functions." *Microbial Ecology in Health and Disease*, 12:193-207. https://doi.org/10.1080/08910600050216183

McFarland, Lynne. 2015. "From yaks to yogurt: the history, development, and current use of probiotics." *Clinical Infectious Diseases*, 60: 85-90. https://doi.org/10.1093/cid/civ054

Meira, Quênia Gramile Silva, Magnani, Marciane, Júnior, Francisco Cesino de Medeiros, Queiroga, Rita de Cássia Ramos do Egito, Madruga, Marta Suely, Gullón, Beatriz, Gomes, Ana Maria Pereira, Pintado, Maria Manuela Estevez and Souza, Evandro Leite de. 2015. "Effects of added *Lactobacillus acidophilus* and *Bifidobacterium lactis* probiotics on the quality characteristics of goat ricotta and their survival under simulated gastrointestinal conditions." *Food Research International,* 76:8282-838. https://doi.org/10.1016/j.foodres.2015.08.002

Metchnikoff, E. 1908. "Optimistic studies New York": *Putman's Sons*, 161-183.

Michael, Minto, Phebus, Randall K. and Schmidt, Karen A. 2010. "Impact of a plant extract on the viability of *Lactobacillus delbrueckii* ssp. *Bulgaricus* and *Streptococcus thermophilus* in nonfat yogurt." *International Dairy Journal*, 20:665-72. https://doi.org/10.1016/j.idairyj.2010.03.005

Min, Min., Bunt, Craig. R.; Mason, Susan L. and Hussain, Malik A. 2019. "Non-dairy probiotic food products: An emerging group of functional foods." *Critical Reviews in*

*Food Science and Nutrition*, 59:2626-41. https://doi.org/10.1080/10408398.2018.1462760

Minelli, Elisa Bertazzoni and Benini, Anna. 2008. "Relationship between several bacteria and their probiotic effects." *Microbial Ecology in Health and Disease*, 20:180-183. https://doi.org/10.1080/08910600802408095

Miranda, Jéssica Soares, Costa, Beatriz Veltre, Oliveira, Isabela Valente de, Lima, Diana Clara Nunes de, Martins, Eliane M. Furtado, Leite Júnior, Bruno Ricardo de Castro, Benevenuto, Wellingta Cristina Almeida do Nascimento, Queiroz, Isabela Campelo de; Silva, Roselir Ribeiro da, Martins, Maurílio Lopes. 2020. "Probiotic jelly candies enriched with native Atlantic Forest fruits and *Bacillus coagulans* GBI-30 6086." *LWT - Food Science and Technology*, 126:109275. https://doi.org/10.1016/j.lwt.2020.109275

Mousa, Haider Abdul-Lateef. 2017. "Prevention and treatment of influenza, influenza-like illness, and common cold by herbal, complementary, and natural therapies." Journal Evidence-Based Complement & Alt Med, 22:166-174. https://doi.org/10.1177/2156587216641831

Mousavi, Malihe, Heshmati, Ali, Garmakhany, Amir Daraei, Vahidinia, A. and Mehdi Taheri. 2019. "Optimization of the viability of *Lactobacillus acidophilus* and physico-chemical, textural and sensorial characteristics of flaxseed-enriched stirred probiotic yogurt by using response surface methodology." *LWT - Food Science and Technology*, 102:80-88. https://doi.org/10.1016/j.lwt.2018.12.023

Nagpal, Ravinder., Kumar, Ashwani, Kumar, Manoj, Behare Pradip V., Jain Shalini and Yadav, Hariom. 2012. "Probiotics, their health benefits and applications for developing healthier foods: A review." *FEMS Microbiology Letters*, 334:1–15. https: //doi.org/10.1111/j.1574-6968.2012.02593.x

Nazir, Yusuf, Hussain, Syed Ammar, Hamid, Aidil Abdul and Song, Y. 2018. "Probiotics and their potential preventive and therapeutic role for cancer, high serum cholesterol, and allergic and HIV diseases." *BioMed Research International*, 29:3428437. https:// doi.org/10.1155/2018/3428437

Nedovic, Viktor, Kalusevic, Ana, Manojlovic, Verica, Levic, Steva and Bugarski, Branko. 2011. "An overview of encapsulation technologies for food applications." Procedia Food Science, 1:1806-1815. https://doi.org/10.1016/j.profoo.2011.09.265

OMS - Organização Mundial da Saúde. *Documentos Básicos*, 39ª ed. Genebra: OMS, (1992).

Paknahad, Zamzam, Moravejolahkami, Amir R. 2021. "Probiotics against viruses; COVID-19 is a paper tiger: a Systematic Review." *Endocrine, Metabolic & Immune Disorders-Drug Targets*, 21:1252-60. https://doi.org/10.2174/18715303206662009 17114033

Parker, R. B.1974. "Probiotics, the other half of the antibiotic story." Animal Nutrition Health, 29:4-8.

Pereira, Eliene Penha Rodrigues, Cavalcanti, Rodrigo Nunes, Esmerino, Erick A., Silva, Ramon, Guerreiro, Lizielle Maria Ricardo, Cunha, R. L., Bolini, Helena Maria André, Meireles, Maria Angela A., Faria, J. A. F. and Cruz, A. G. 2016. "Effect of incorporation of antioxidants on the chemical, rheological, and sensory properties of probiotic petit suisse cheese." *Journal of Dairy Science*, 99:1762-72. https://doi.org/10.3168/jds.2015-9701

Peres, Cátia M., Peres, Cidália, Hernández-Mendoza, Adrián, Malcata, F. Xavier. 2012. "Review on fermented plant materials as carriers and sources of potentially probiotic lactic acid bacteria -With an emphasis on table olives." *Trends in Food Science & Technology*, 26:31-42. https://doi.org/10.1016/j.tifs.2012.01.006

Pinto, Teresa and Vilela, Alice. "Healthy drinks with lovely colors: phenolic compounds as constituents of functional beverages." *Beverages*, 7 (2021):12. https://doi.org/10.3390/beverages7010012

Reid, Gregor. 2016. "Probiotics: definition, scope, and mechanisms of action." Best Practice, 30:17-25. https://doi.org/10.1016/j.bpg.2015.12.001

Reid, Gregor. 2015. "The growth potential for dairy probiotics." *International Dairy Journal,* 49:16-22. https://doi.org/10.1016/j.idairyj.2015.04.004

Reid, Gregor, Gadir, Azza A. and Dhir, Raja. 2019. "Probiotics: reiterating what they are and what they are not." *Frontiers Microbiology*, 10:424. https://doi.org/ 10.3389/fmicb.2019.00424

Rembacken, B. J., Snelling, A. M., Hawkey, P. M., Chalmers, D. M. and Axon, A. T. 1999. "Nonpathogenic *Escherichia coli* versus mesalazine for the treatment of ulcerative colitis: A randomized trial." *Lancet*, 354:635-639.

Rettger, L. F. and Cheplin, H. A. 1922. "*Bacillus acidophilus* and its therapeutic application." *Archives of Internal Medicine,* 29:357-367.

Rolf, Rial D. 2000. "The role of probiotic cultures in the control of gastrointestinal health." *The Journal of Nutrition*, 130:396S-402S. https://doi.org/10.1093/jn/130.2.396S

Ruiz, Lorena, Delgado, Susana, Ruas-Madiedo, Patricia, Margolles, Abelardo and Sánchez, Borja. 2016. "Proteinaceous molecules mediating *Bifidobacterium*-host interactions." *Frontiers in Microbiology*, 7:1193. https://doi.org/10.3389/fmicb.2016.01193

Sanders, Mary Ellen. 2008. "Probiotics: Definition, Sources, Selection, and Uses." Clinical Infectious Diseases, 46:58-61. https://doi.org/10.1086/523341

Shruthi, Balakrishna, Deepa, Nagaraju, Somashekaraiah, Rakesh, Adithi, G., Divyashree, S. and Sreenivasa, Marikunte Yanjarappa. 2022. "Exploring biotechnological and functional characteristics of probiotic yeasts: A review." *Biotechnology Reports*, 34:e00716. https://doi.org/10.1016/j.btre.2022.e00716

Silva, Thamires Maria Simões da, Piazentin, Anna Carolina Meirelles, Mendonça, Carlos Miguel Nóbrega, Converti, Attilio, Bogsan, Cristina Stewart Bittencourt, Mora, Diego and Oliveira, Ricardo Pinheiro de Souza. 2020. "Buffalo milk increases viability and resistance of probiotic bacteria in dairy beverages under in vitro simulated gastrointestinal conditions." *Journal of Dairy Science*, 103:7890-7897.https://doi.org/10.3168/jds.2019-18078

Simpson, Paul J., Stanton, Catherine, Fitzgerald, Gerald F. and Ross, R. P. 2005. "Intrinsic tolerance of Bifidobacterium species to heat and oxygen and survival following spray drying and storage." *Journal of Applied Microbiology*, 99:493-501. https://doi.org/10.1111/j.1365-2672.2005.02648.x

Solanki, Himanshu K., Pawar, Dipak D., Shah, Dushyant A., Prajapati, Vipul D., Jani, Girish K., Mulla, Akil M. and Thakar, Prachi M. 2013. "Development of microencapsulation delivery system for long-term preservation of probiotics as biotherapeutics agent." *BioMed Research International*, 2013:1-21. https://doi.org/10.1155/2013/620719

Sornsenee, Phoomjai, Chimplee, Siriphorn, Saengsuwan, Phanvasri and Romyasamit, Chonticha. 2022. "Characterization of probiotic properties and development of banana powder enriched with freeze-dried *Lacticaseibacillus paracasei* probiotics." *Heliyon*, 8: e11063. https://doi.org/10.1016/j.heliyon.2022.e11063

Stadnik, Joanna, Dolatowski and Zbigniew J. 2014. "Effect of inoculation with probiotics and ageing time on selected functional properties and oxidation of proteins in dry-cured pork loins." *International Journal Food Properties,* 17:866–76. https://doi.org/10.1080/10942912.2012.685679

Storz, Gisela and Imlay, James A. 1999. "Oxidative stress." *Current Opinion Microbiology*, 2:188-194. https://doi.org/10.1016/S1369-5274(99)80033-2

Szymański, H., Pejcz, J., Jawień, M., Chmielarczyk, A., Strus, M., and Heczko, P. B. 2006. "Treatment of acute infectious diarrhoea in infants and children with a mixture of three *Lactobacillus rhamnosus* strains—A randomized, double-blind, placebo-controlled trial." *Alimentary Pharmacology & Therapeutics*, 23:247-53. https://doi.org/10.1111/j.1365-2036.2006.02740.x

Talwalkar, Akshat and Kailasapathy, Kasipathy. 2004. "A review of oxygen toxicity in probiotic yogurts: Influence on the survival of probiotic bacteria and protective techniques." *Comprehensive Reviews in Food Science and Food Safety*, 3:117-124. https://doi.org/10.1111/j.1541-4337.2004.tb00061.x

Tegene, Bantayehu Addis and Kebede, Bekalu. 2022. "Probiotics, their prophylactic and therapeutic applications in human health development: A review of the literature." *Helyon,* 8:e09725. https://doi.org/10.1016/j.heliyon.2022.e09725

Tissier H. 1906. "Traitment de infections intestinales per la methode de la flore bacterienne de I'intestin." [Treatment of intestinal infections by the method of the bacterial flora of the intestine].

Tomar, Oktay. 2019. "The effects of probiotic cultures on the organic acid content, texture profile and sensory attributes of Tulum cheese." *International Journal Dairy Technology*, 72:218-228. https://doi.org/10.1111/1471-0307.12574

Tripathi, Manoj Kumar and Giri, Saroj Kumar. 2014. "Probiotic functional foods: Survival of probiotics during processing and storage." *Journal of Functional Foods*, 9:225-241. https://doi.org/10.1016/j.jff.2014.04.030

Usaga, Jessie, Barahona, Daniela, Arroyo Laura, Esquivel, Patricia. 2022. "Probiotics survival and betalains stability in purple pitaya (*Hylocereus* sp.) juice." *NFS Journal*, 27: 47-53. https://doi.org/10.1016/j.nfs.2022.05.001

Valencia, Marcela Sarmento, Salgado, Silvana Magalhães, Andrade, Samara Alvachian Cardoso Andrade, Padilha, Vivianne Montarroyos, Livera, Alda Verônica Souza and Stamford, Tânia Lúcia Montenegro. 2016. "Development of creamy milk chocolate dessert added with fructooligosaccharide and *Lactobacillus paracasei* subsp. *paracasei* LBC 81." *LWT - Food Science and Technology*, 69:104-109. https://doi.org/10.1016/j.lwt.2016.01.039

Vasudha, Sharma and Mishra, Hari N. 2013. "Non dairy probiotic beverages." *International Food Research*, 20:7–15.

Verschuere, Laurent, Rombaut, Geert, Sorgeloos, Patrick and Verstraete Willy. 2000. "Probiotic bacteria as biological control agents in aquaculture." *Microbiology*, 64:655-671. https://doi.org/10.1128/MMBR.64.4.655-671.2000

Vivek, Kambhampati, Mishra, Sabyasachi, Pradhan, Rama Chandra, Nagarajan, Meena, Kumar, Pavitra Krishna, Singh, Singam Suranjoy, Manvi, Dronachari and Gowda, Nanje Na. 2023. "A comprehensive review on microencapsulation of probiotics: technology, carriers and current trends." *Applied Food Research*, 3:100248. https://doi.org/10.1016/j.afres.2022.100248

Wang, Zining, Feng, Yunzi, Yang, Nana, Jiang, Tian, Xu, Huaide and Lei Hongjie. 2022. "Fermentation of kiwifruit juice from two cultivars by probiotic bacteria: Bioactive phenolics, antioxidant activities, and flavor volatiles." *Food Chemistry*, 373: 131455. https://doi.org/10.1016/j.foodchem.2021.131455

Wessels, Stephen, Axelsson, Lars, Hansen, Egon Bech, Vuyst, Luc De, Laulund, Svend, LähteenmäkI, Liisa, Lindgren, Sven, Mollet, Beat, Salminen, Seppo, and Wright, von Atte. 2004. "The lactic acid bacteria, the food chain, and their regulation." *Trends in Food Science and Technology*, 15:498-505. https://doi.org/10.1016/j.tifs.2004.03.003

Wilkins, T., Sequoia, Jacqueline. 2017. "Probiotics for gastrointestinal conditions: A summary of the evidence." *American Family Physician*, 96:170-178.

Xu, Cong, Ban, Qingfeng, Wang, Wang, Hou, Juncai and Jiang Zhanmei. 2022. "Novel nano-encapsulated probiotic agents: Encapsulate materials, delivery, and encapsulation systems." *Journal of Controlled Release*, 349:184-205. https://doi.org/10.1016/j.jconrel.2022.06.061

Xu, Zhe, Shi, Lei, Wang, Yijin, Zhang, Jiyuan, Huang, Lei, Zhang, Chao, Liu, Shuhong, Zhao, Peng, Liu, Hongxia, Zhu Li, Tai, Yanhong, Bai Changing, Gao, Tinting, Song, Jinwen, Xia, Peng, Dong, Jinghui, Zhao, Jingmin, Zhao and Wang Fu-Sheng. 2020. "Pathological findings of COVID-19 associated with acute respiratory distress syndrome." *The Lancet Respiratory Medicine*, 8:420-422. https://doi.org/10.1016/S2213-2600(20)30076-X

Yadav, A., Chandra, H., Maurya, V. K. 2017. "Probiotics: recent advances and prospects." *Journal of Plant Development Sciences*, 9:967-975.

Yeo, Siok-Koon, Liong, Min-Tze. 2010. "Angiotensin I-converting enzyme inhibitory activity and bioconversion of isoflavones by probiotics in soymilk supplemented with prebiotics." *International Journal of Food Sciences and Nutrition*, 61:161–81. https://doi.org/10.3109/09637480903348122

Zahrani, Ashwag Jaman AL and Shori, Amal Bakr. 2023."Viability of probiotics and antioxidant activity of soy and almond milk fermented with selected strains of probiotic *Lactobacillus spp*." *LWT - Food Science and Technology*, 176: 114531. https://doi.org/10.1016/j.lwt.2023.114531

Zoellner, S. S., Cruz, A. G., Faria, J. A. F., Bolini, Helena, Moura, Mirian Ribeiro Leite Moura, Carvalho, Lucia Maria Jaeger de and Sant'Ana, A. S. 2009. "Whey beverage with acai pulp as a food carrier of probiotic bacteria." *Australian Journal of Dairy Technology*, 64:177-181.

Zubillaga, Marcela, Weill, Ricardo, Postaire, Eric, Goldman, Cinthia, Caro, Ricardo and Boccio, José. 2001."Effect of probiotics and functional foods and their use in different diseases." *Nutrition Research*, 21:569-79. https://doi.org/10.1016/S0271-5317(01)002810

# Chapter 2

# The History of Microencapsulation: The Path Towards Probiotic Protection

**Karen Cristina Guedes Silva[1,*], PhD and Flávia Souza Almeida[2], PhD**

[1]Federal Institute of Paraná – IFPR - Campus Assis Chateaubriand, Assis Chateaubriand, Paraná, Brazil

[2]Department of Food Engineering and Technology, School of Food Engineering, University of Campinas – UNICAMP, Campinas, São Paulo, Brazil

## Abstract

Microencapsulation has been known as a process used to protect bioactive compounds inside polymeric matrices. The technique is used for different purposes such as in the protection of the core material from adverse conditions, in the promotion of controlled delivery, and for preserving and/or masking of flavors. Currently, several techniques have been used to produce encapsulation systems, but to achieve mastery of these techniques, pioneering studies were fundamental for the development of the microencapsulation area by science. One of the first reports in the literature associated with encapsulation dates back to 1872. It describes a process for obtaining powders by spray drying, marking the beginning of the development of one of the most used techniques for encapsulation. A brief history of the first reports that allowed the development of other encapsulation techniques will be presented in this chapter such as electrospray, coacervation, liposomes, fluidization, and layer-by-layer, among others. In the development of these studies, several wall materials were evaluated, most of them explored biopolymers that are still used today, namely gelatin, carboxymethylcellulose, starch, zein,

* Corresponding Author's Email: karen.silva@ifpr.edu.br.

In: Microencapsulation of Probiotics
Editors: D. de Almeida Paula, K. Cristina Guedes Silva and N. de Almeida Costa
ISBN: 979-8-88697-979-4

and waxes. Among the discoveries of materials sensitive to adverse conditions, probiotics gain attention due to their potential to promote health benefits. The evolution of different encapsulation techniques to improve the viability of probiotics, enabling both their protection and controlled release, in addition to the application of encapsulated probiotic cells in food products, pharmaceuticals, and supplements.

**Keywords**: probiotics, history, microencapsulation, coating techniques

## Introduction

Microencapsulation can be defined as a process that allows active core materials to be coated or incorporated into a polymeric matrix. This technique provides a physical barrier to the trapped material, which may be referred to as a core or internal phase, and the protective structure may be referred to as the membrane or wall material (Gharsallaoui et al., 2007). The use of microencapsulation in foods provides numerous advantages such as the protection of sensitive compounds to avoid degradation by light, oxygen, humidity, acidity, and reactive ingredients. Moreover, it minimizes nutritional losses, incorporates controlled release mechanisms, masks, or preserves flavors and aromas, besides preventing crowds (Shahidi and Han 1993). In this context, a brief history of studies and techniques that allowed the development of microencapsulation in science and its application to improve probiotc's viability and resistance to adverse conditions will be discussed in the following paragraphs.

## History of Microencapsulation

One of the first reports in literature associated with encapsulation dates back to 1872, in a patent signed by Percy in New York (Percy 1872). His discovery dealt with the description of the drying process by spray drying, where fluid substances could be atomized and dried simultaneously by passing through a tube. At the exit, they were atomized and exposed to currents of air or other gases that could be at low humidity, hot or cold. These process conditions allowed the water present in the fluid to evaporate and produce fine powders, presenting varied particle sizes according to the conditions adopted. In this aforementioned patent, materials such as dextrin and gelatin are already introduced, besides the fact that the technique dried materials quickly and reduced destructive chemical changes. Following this creation, this process could be applied a few years later in the manufacture of encapsulation systems, despite its first report only addressing the drying of fluids for the production of fine powders.

In 1834, gelatin-based capsules were described in the Bulletin of the Laws of the Kingdom of France, by Joseph Gerard Dublanc and François Achille Barnabe Mothes. The invention was based on the development of gelatin capsules for application in drugs and, during the fabrication, the capsules needed to be diluted in oil or other liquids to prevent their compaction into solid materials (Rana et al., 2022). It is noteworthy that the manufacture of gelatin capsules presented many issues along the process, such as leaks and air bubbles, and, consequently, caused problems of standardization in the number of medicines produced.

To overcome the leakage and formation of air bubbles, in 1931, Scherer improved the technique to form coverage on fluid substances, allowing an automatic operation without the need for manipulation. The equipment consisted of feeding the coating material (described as gelatin in its patent) in a system of plates that converged. The fluid was inserted into an internal cavity of the material occupying one of the feeding directions, which later overlapped and sealed it (Scherer 1931).

The first experiment related to electrospray systems was reported in 1917, where a liquid disintegrated due to instability caused by the action of an electric field. The experiment was observed under a microscope and photographed. Zeleny (1917) noticed the formation of elongated drops of ethyl alcohol when subjected to a charge in the order of 5000 volts or more. The electric field promoted an instability that allowed the formation of a hanging thread of liquid that formed small drops when broken.

Encapsulation systems on a microscopic scale were produced for the first time in 1957, with the invention of Green and Schleicher, who explored the formation of complexes by chemical interaction between ionizable and oppositely charged components, named coacervation (Green and Schleicher 1957; Sobel, Versic, and Gaonkar 2014). In this technique, envelopes on oil droplets were formed after dilution and/or adjustment of the pH of the complex colloid material mixture (gum Arabic and gelatin). The finished product may be used for coatings or films and since then, this process has been improved and used for encapsulation of several bioactive components, especially probiotic microorganisms (Silva et al., 2021; Paula et al., 2019; Qi et al., 2021).

A few years later, the idea of chemical interaction between opposite-charged compounds used for coacervation became the basis for the development of particles produced with a magnetic material such as iron oxide, cobalt, and nickel, among others, involving oils as the core material. These capsules ranged in size from 3 to 150 μm and could be ruptured by pressure. In a patent registered in 1961, Schleicher and Baughman proposed its application in printers by magnetic means, in which a colorless ink was encapsulated and applied over a layer of paper. After the rupture of the capsules through pressure, the ink was released and, when in contact with the reagent that was also incorporated into the paper, gave rise to color (Schleicher and Baughman 1961).

Liposomal cell-wall-like structures, in other words, concentric bilayers of phospholipids that spontaneously assemble, were first observed in an exploratory study with phospholipids around 1960 (Bangham, A. D., Horne, R. W., 1962 apud Sobel, Versic, and Gaonkar 2014 p. 6). The combination of hydrophilic and hydrophobic compounds was also proposed by Schneider in 1974 (Schneider 1974) as an improved method for the development of liposomes for encapsulation. The idea of proposed method for liposome fabrication consisted of dispersing an aqueous liquid together with compounds containing a polar hydrophilic and a non-polar hydrophobic group to form a colloidal dispersion, which was subsequently subjected to an ultrasonic frequency vibration. Liposome encapsulation became a well-known method from the 1970s onwards, but only after the 90's, studies started giving notoriety to the technique (Timilsena, Haque, and Adhikari 2020).

In 1965, Wurster and Lindlof registered a patent containing an apparatus for the encapsulation of discrete particles of macro-size, giving rise to studies with fluidized beds. Compounds such as sugar syrup, gelatin, carboxymethylcellulose, methylcellulose, starch, zein, and waxes were listed as possible encapsulation materials, besides emulsions. The apparatus consisted of a system that allowed the core material to get in contact with a zone of finely atomized coating droplets, while they remained suspended in a current of air. Under a fluidized condition countercurrent to a flow of heated gas, the dried coated particles were able to descend the chamber and reenter the coating portion of the cycle (Wurster and Lindlof 1965).

Considering the growing interest and progress in the encapsulation field, in 1982, Sozzi and coauthors patented the process for producing microcapsules to protect microorganisms in an attempt to replace antibiotics with probiotic cultures capable of recomposing the intestinal flora. The microcapsules were designed to protect microorganisms of the genus

*Lactobacillus* and *Bifidobacterium*, as the particles presented anaerobiosis characteristics and prolonged the microorganism's viability. The authors developed microcapsules by dispersing the cell culture in melted fat, followed by spraying in a chamber with liquid nitrogen, air, or another cold medium, indicating the possibility of adding a protein such as zein to reduce agglomeration. Particles from 0.1 to 5 mm, and resistant to environmental and stomach conditions were obtained (Sozzi, Schrenk, and Buhler 1982).

To increase the viability of probiotic microorganisms, a dry granular product was further prepared by fluidization. The encapsulation system prevented oxygen and moisture permeability, increasing the survival of microorganism cells compared to the conventional product during prolonged storage periods. For that, food-grade biopolymers were proposed as coating materials (Okonogi, Tomita, and Shimamura 1989). Some years later, the layer-by-layer method emerged in the 1990s (Decher, Hong, and Schmitt 1992). The first report in the literature showed that multilayer films composed of at least 100 alternating layers could be constructed by cyclically alternating the immersion of solid support in anionic and cationic solutions. The technique made it possible to improve the protection and controlled delivery of sensitive compounds, such as probiotics (Cheng et al., 2021; Chen et al., 2023).

In 1989, the concept of probiotics was proposed by Fuller, as a "Live microbial feed supplement which beneficially affects the host animal by improving its intestinal microbial balance" (Fuller 1989). A few years later (1907), lactic acid bacteria were also studied by Elie Metchnikoff, a Nobel Prize-winning Russian scientist, who also assumed their potential to promote health benefits (Rodrigues et al., 2020).

Currently, international agencies such as the Food and Agriculture Organization of the United Nations (FAO) and the World Health Organization (WHO) define probiotics as "live microorganisms which when administered in adequate amounts confer a health benefit on the host" (FAO and WHO 2006).

After these discoveries related to the probiotics potential over the years, many researchers have investigated their action in promoting health, such as in the formation or reconstruction of intestinal microflora, reduction in serum cholesterol levels, a decrease of lactose intolerance, improvement of calcium absorption, inactivation or inhibition of carcinogenic compounds, anti-diabetic properties, support of immuno-compromised host and pathogens combat (Wood 2012; Bagher Hashemi et al., 2022; Gomes and Malcata 1999; Havenaar and Huis In't Veld 1992; Saito et al., 1980; Zhong et al., 2021), besides the reduction of the risk associated with cardiovascular diseases

(Agerholm-Larsen et al., 2000; Oniszczuk et al., 2021), the improvement of the anti-hypercholesterolemia and anti-hyperlipidemic effect (Li et al., 2020; Taranto et al., 1998; Tian et al., 2022), among other conditions.

In this sense, the interest in incorporating probiotics into foods for health maintenance has resulted in the development of various products (Chugh and Kamal-Eldin 2020; Rodrigues et al., 2020; Terpou et al., 2019), such as fermented dairy (Zepeda-Hernández et al., 2021) and non-dairy products (Kumar et al., 2022), which have been improved with different encapsulation techniques since the discovery of the potential of probiotics, as well as their sensitivity to adverse conditions (Rodrigues et al., 2020).

The evolution of encapsulation systems and a better understanding of the probiotic mechanisms of action for health promotion enabled the development of many microencapsulation techniques for protecting probiotic cells. For example, it can be mentioned varied and updated studies using the technique of extrusion (How et al., 2022), emulsion (Camelo-Silva et al., 2022), spray drying (Sharma, Rashidinejad, and Jafari 2022), spray chilling (Silva et al., 2022), fluidized bed (Mirzamani et al., 2021), electrospray (Zaeim et al., 2019), and many others.

## Conclusion

To arrive at the current knowledge about encapsulation systems and employ different techniques according to desired applications, it is important to emphasize the role of history. The first works and discoveries enabled explorations that culminated in the development of the encapsulation field.

Currently, encapsulation technology has been employed by industries for diverse applications. Encapsulation has become increasingly present in the development of pharmaceuticals, nutraceuticals, and functional food industries, presenting potential as a highly effective method to increase stability, ensure protection, and controlled delivery of compounds of interest (Timilsena, Haque, and Adhikari 2020). In the subsequent chapters, technologies and processes related to probiotics encapsulation will be described in more detail, as well as the materials used and food applications.

## References

Agerholm-Larsen, L., A. Raben, N. Haulrik, A. S. Hansen, M. Manders, and A. Astrup. 2000. "Effect of 8 Week Intake of Probiotic Milk Products on Risk Factors for Cardiovascular Diseases." *European Journal of Clinical Nutrition* 54 (4): 288–97. https://doi.org/10.1038/sj.ejcn.1600937.

Bagher Hashemi, Seyed Mohammad, Elahe Abedi, Shima Kaveh, and Maryam Mousavifard. 2022. "Hypocholesterolemic, Antidiabetic and Bioactive Properties of Ultrasound-Stimulated Exopolysaccharide Produced by Lactiplantibacillus Plantarum Strains." *Bioactive Carbohydrates and Dietary Fibre* 28: 100334. https://doi.org/10.1016/j.bcdf.2022.100334.

Camelo-Silva, C., Silvani Verruck, Alan Ambrosi, and Marco Di Luccio. 2022. "Innovation and Trends in Probiotic Microencapsulation by Emulsification Techniques." *Food Engineering Reviews* 14 (3): 462–90. https://doi.org/10.1007/s12393-022-09315-1.

Chen, X., Ke Zhang, Yan'an Wei, Yongyong Wang, Xiaofan Wu, and Haifeng Wan. 2023. "Green Tea Polysaccharide Conjugates and Gelatin Enhanced Viability of L. Acidophilus by Layer-by-Layer Encapsulation." *Food Bioscience* 52: 102471. https://doi.org/10.1016/j.fbio.2023.102471.

Cheng, Q., Lu Liu, Mingzhi Xie, Hang Li, Dong Ma, and Wei Xue. 2021. "A Colon-Targeted Oral Probiotics Delivery System Using an Enzyme-Triggered Fuse-Like Microcapsule." *Advanced Healthcare Materials* 10 (8): 2001953. https://doi.org/10.1002/adhm.202001953.

Chugh, B., and Afaf Kamal-Eldin. 2020. "Bioactive Compounds Produced by Probiotics in Food Products." *Current Opinion in Food Science* 32: 76–82. https://doi.org/10.1016/j.cofs.202002.003

Decher, G., J. D. Hong, and J. Schmitt. 1992. "Buildup of Ultrathin Multilayer Films by a Self-Assembly Process: III. Consecutively Alternating Adsorption of Anionic and Cationic Polyelectrolytes on Charged Surfaces." *Thin Solid Films* 210–211: 831–35. https://doi.org/10.1016/0040-6090(92)90417-A.

FAO, Food and Agriculture Organization of the United States, and World Health Organization WHO. 2006. "Probiotics in Food: Health and Nutritional Properties and Guidelines for Evaluation." https://www.fao.org/3/a0512e/a0512e.pdf.

Fuller, R. AFRC. 1989. "Probiotics in Man and Animals." *Journal of Applied Bacteriology* 66 (5): 365–78. https://doi.org/10.1111/j.1365-2672.1989.tb05105.x.

Gharsallaoui, A., Gaëlle Roudaut, Odile Chambin, Andrée Voilley, and Rémi Saurel. 2007. "Applications of Spray-Drying in Microencapsulation of Food Ingredients: An Overview." *Food Research International* 40 (9): 1107–21. https://doi.org/10.1016/j.foodres.2007.07.004

Gomes, A. M. P., and F. Xavier Malcata. 1999. "Bifidobacterium Spp. and Lactobacillus Acidophilus: Biological, Biochemical, Technological and Therapeutical Properties Relevant for Use as Probiotics." *Trends in Food Science & Technology* 10 (4): 139–57. https://doi.org/10.1016/S0924-2244(99)00033-3.

Green, K. Barrett, and Lowe Schleicher. 1957. *Oil containing microscopic capsules and methods of making them.* No. 365,105, issued 1957.

https://patentimages.storage.googleapis.com/39/a9/25/208f85dd7fdbda/US2800457.pdf

Havenaar, R., and Jos H. J. Huis In't Veld. 1992. "*Probiotics: A General View BT - The Lactic Acid Bacteria Volume 1: The Lactic Acid Bacteria in Health and Disease*." In, edited by Brian J. B. Wood, 151–70. Boston, MA: Springer US. https://doi.org/10.1007/978-1-4615-3522-5_6

How, Y. H., Ka-Wai Lai, Liew-Phing Pui, and Lionel Lian-Aun In. 2022. "Co-Extrusion and Extrusion Microencapsulation: Effect on Microencapsulation Efficiency, Survivability through Gastrointestinal Digestion and Storage." *Journal of Food Process Engineering* 45 (3): e13985. https://doi.org/10.1111/jfpe.13985.

Kumar, S., Gurdeep Rattu, Swati Mitharwal, Abhishek Chandra, Sourabh Kumar, Aman Kaushik, Vijendra Mishra, and Prabhat K Nema. 2022. "Trends in Non-Dairy-Based Probiotic Food Products: Advances and Challenges." *Journal of Food Processing and Preservation* 46 (9): e16578. https://doi.org/10.1111/jfpp.16578.

Li, X., Yuchun Xiao, Liqiong Song, Yuanming Huang, Qiongfang Chu, Siyi Zhu, Shan Lu, Hou, Luwen, Li, Zhen, Li, Jianguo, Xu, Jianguo, Ren, Zhihong. 2020. "Effect of *Lactobacillus Plantarum* HT121 on Serum Lipid Profile, Gut Microbiota, and Liver Transcriptome and Metabolomics in a High-Cholesterol Diet–Induced Hypercholesterolemia Rat Model." *Nutrition* 79–80: 110966. https://doi.org/10.1016/j.nut.2020.110966

Mirzamani, S. S., A. R. Bassiri, H. Tavakolipour, M. H. Azizi, and M. Kargozari. 2021. "Survival of Fluidized Bed Encapsulated Lactobacillus Acidophilus under Simulated Gastro-Intestinal Conditions and Heat Treatment during Bread Baking." *Journal of Food Measurement and Characterization* 15 (6): 5477–84. https://doi.org/10.1007/s11694-021-01108-0

Okonogi, O. S., Mamoru Tomita, and S. Shimamura. 1989. Granular Product Of Dried Microorganism Cells And Manufacturing Method Therefor. 4.888.171, issued 1989. https://patentimages.storage.googleapis.com/7d/bc/f1/718c9ed76c534e/US4888171.pdf.

Oniszczuk, A., Tomasz Oniszczuk, Marek Gancarz, and Jolanta Szymańska. 2021. "Role of Gut Microbiota, Probiotics, and Prebiotics in the Cardiovascular Diseases." *Molecules*. https://doi.org/10.3390/molecules26041172.

Paula, D. de A., Martins, E. M. F., Costa, N. de A, Oliveira, P. M. de, Oliveira, E. B. de and Ramos, A. M. 2019. "Use of Gelatin and Gum Arabic for Microencapsulation of Probiotic Cells from Lactobacillus Plantarum by a Dual Process Combining Double Emulsification Followed by Complex Coacervation." *International Journal of Biological Macromolecules* 133: 722–31. https://doi.org/10.1016/j.ijbiomac.2019.04.110

Percy, S. R. 1872. *Improvement in drying and concentrating liquid substances by atomizing.* No. 125,406, issued 1872. https://patentimages.storage.googleapis.com/ee/16/8d/7262ae44781662/US125406.pdf

Qi, X., Yang Lan, Jae-Bom Ohm, Bingcan Chen, and Jiajia Rao. 2021. "The Viability of Complex Coacervate Encapsulated Probiotics during Simulated Sequential

Gastrointestinal Digestion Affected by Wall Materials and Drying Methods." *Food & Function* 12 (19): 8907–19. https://doi.org/10.1039/D1FO01533H.

Rana, D., Sagar Salave, Garima Rawat, and Derajram Benival. 2022. "Recent Trends in Drug Delivery and Emerging Biomedical Applications of Gelatin for Ophthalmic Indications." *Macromolecular Research* 30 (10): 687–702. https://doi.org/10.1007/s13233-022-0078-9

Rodrigues, F. J., M. F. Cedran, J. L. Bicas, and H. H. Sato. 2020. "Encapsulated Probiotic Cells: Relevant Techniques, Natural Sources as Encapsulating Materials and Food Applications – A Narrative Review." *Food Research International* 137: 109682. https://doi.org/10.1016/j.foodres.2020.109682.

Saito, H., T. Watanabe, Y. Horikawa, and O. Tado. 1980. "Protective Effect of Lactobacilli on Experimental Escherichia Coli Infection." *Medicine and Biology, Japan* 101 (1): 61–64.

Scherer, R. P. 1931. Method of machine for making capsules. No. 588,230, issued 1931. https://patentimages.storage.googleapis.com/9d/d4/fd/4f26d776cce4fc/US1970396.pdf.

Schleicher, L., and Charles S. Baughman. 1961. Macroscopic capsules containing magnetizable material. No. 637,270, issued 1961. https://patentimages.storage.googleapis.com/d9/44/d6/07b1e71723d4bb/US2971916.pdf.

Schneider, M. 1974. *Method For Producing Liposomes.* De-2532317-A1, issued 1974.

Shahidi, F., and Xiao-Qing Han. 1993. "Encapsulation of Food Ingredients." *Critical Reviews in Food Science and Nutrition* 33 (6): 501–47. https://doi.org/10.1080/10408399309527645

Sharma, R., Ali Rashidinejad, and Seid Mahdi Jafari. 2022. "Application of Spray Dried Encapsulated Probiotics in Functional Food Formulations." *Food and Bioprocess Technology* 15 (10): 2135–54. https://doi.org/10.1007/s11947-022-02803-6.

Silva, R., Tatiana Colombo Pimentel, Fernando Eustáquio de Matos Junior, Erick Almeida Esmerino, Monica Q Freitas, Carmen Silvia Fávaro-Trindade, Márcia Cristina Silva, and Adriano Gomes Cruz. 2022. "Microencapsulation with Spray-Chilling as an Innovative Strategy for Probiotic Low Sodium Requeijão Cremoso Processed Cheese Processing." *Food Bioscience* 46: 101517. https://doi.org/10.1016/j.fbio.2021.101517

Silva, T. M. da, Vandré Sonza Pinto, Vítor Ramires Fonseca Soares, Débora Marotz, Alexandre José Cichoski, Leila Queiroz Zepka, Eduardo Jacob Lopes, Cristiane de Bona da Silva, and Cristiano Ragagnin de Menezes. 2021. "Viability of Microencapsulated Lactobacillus Acidophilus by Complex Coacervation Associated with Enzymatic Crosslinking under Application in Different Fruit Juices." *Food Research International* 141: 110190. https://doi.org/10.1016/j.foodres.2021.110190

Sobel, R., Ronald Versic, and Anilkumar G Gaonkar. 2014. "Chapter 1 - Introduction to Microencapsulation and Controlled Delivery in Foods." In, edited by Anilkumar G Gaonkar, Niraj Vasisht, Atul Ramesh Khare, and Robert B. *T - Microencapsulation in the Food Industry Sobel,* 3–12. San Diego: Academic Press. https://doi.org/10.1016/B978-0-12-404568-2.00001-7

Sozzi, L. T., Orbe A. Schrenk, and Tolochenaz M. Buhler. 1982. *Microcapsule containing a microorganism and a process for its production.* No.: 96,771, issued 1982.

Taranto, M. P., M. Medici, G. Perdigon, A. P. Ruiz Holgado, and G F Valdez. 1998. "Evidence for Hypocholesterolemic Effect of Lactobacillus Reuteri in Hypercholesterolemic Mice." *Journal of Dairy Science* 81 (9): 2336–40. https://doi.org/10.3168/jds.S0022-0302(98)70123-7

Terpou, A., Aikaterini Papadaki, Iliada K Lappa, Vasiliki Kachrimanidou, Loulouda A Bosnea, and Nikolaos Kopsahelis. 2019. "Probiotics in Food Systems: Significance and Emerging Strategies Towards Improved Viability and Delivery of Enhanced Beneficial Value." *Nutrients*. https://doi.org/10.3390/nu11071591

Tian, L., Rongmei Liu, Zhiwei Zhou, Xiaofang Xu, Su Feng, Ariel Kushmaro, Robert S Marks, Dan Wang, and Qun Sun. 2022. "Probiotic Characteristics of Lactiplantibacillus Plantarum N-1 and Its Cholesterol-Lowering Effect in Hypercholesterolemic Rats." *Probiotics and Antimicrobial Proteins* 14 (2): 337–48. https://doi.org/10.1007/s12602-021-09886-1

Timilsena, Y. P., Md Amdadul Haque, and Benu Adhikari. 2020. "Encapsulation in the Food Industry: A Brief Historical Overview to Recent Developments." *Food and Nutrition Sciences* 11 (06): 481. https://doi.org/10.4236/fns.2020.116035

Wood, B. J. B. 2012. *The Lactic Acid Bacteria: Volume 1: The Lactic Acid Bacteria in Health and Disease*. Springer Science & Business Media.

Wurster, D. E., and James A. Lindlof. 1965. Apparatus for the encapsulation of discrete particles. US3196827A, issued 1965. https://patents.google.com/patent/US3196827

Zaeim, D., Mahboobe Sarabi-Jamab, Behrouz Ghorani, and Rassoul Kadkhodaee. 2019. "Double Layer Co-Encapsulation of Probiotics and Prebiotics by Electro-Hydrodynamic Atomization." *LWT* 110: 102–9. https://doi.org/10.1016/j.lwt.2019.04.040

Zeleny, J. 1917. "Instability of Electrified Liquid Surfaces." *Physical Review* 10 (1): 1–6. https://doi.org/10.1103/PhysRev.10.1

Zepeda-Hernández, A., Luis Eduardo Garcia-Amezquita, Teresa Requena, and Tomás García-Cayuela. 2021. "Probiotics, Prebiotics, and Synbiotics Added to Dairy Products: Uses and Applications to Manage Type 2 Diabetes." *Food Research International* 142: 110208. https://doi.org/10.1016/j.foodres.2021.110208

Zhong, H., Abdullah, Yipeng Zhang, Minjie Zhao, Junhui Zhang, Hui Zhang, Yuhang Xi, Haiying Cai, and Fengqin Feng. 2021. "Screening of Novel Potential Antidiabetic Lactobacillus Plantarum Strains Based on *in Vitro* and *in Vivo* Investigations." *LWT* 139: 110526. https://doi.org/10.1016/j.lwt.2020.110526

# Chapter 3

# Coating Materials Used in the Encapsulation of Probiotics

**Carini Aparecida Lelis[1,*], PhD**
**Rafaela Teixeira Rodrigues do Vale[2], PhD**
**and Carlos Adam Conte, Jr.[1,3], PhD**
[1]Center for Food Analysis (NAL), Technological Development Support Laboratory (LADETEC), Federal University of Rio de Janeiro (UFRJ), Rio de Janeiro, Brazil
[2]Minas Gerais Agricultural Research Company (EPAMIG), Cândido Tostes Dairy Institute (ILCT), Santa Terezinha, Juiz de Fora, Minas Gerais, Brazil
[3]Laboratory of Advanced Analysis in Biochemistry and Molecular Biology (LAABBM), Department of Biochemistry, Federal University of Rio de Janeiro (UFRJ), Rio de Janeiro, Brazil

## Abstract

The resistance of probiotics when passing through the gastrointestinal system and subsequent absorption is essential for obtaining possible health benefits. Many probiotics are added to foods to be ingested during eating. However, the viability of probiotics during food processing and storage is also an important factor. Protective mechanisms for probiotic cells have been developed over the years. The encapsulation of probiotic cells can occur through different techniques and using different encapsulation materials. The most used materials are of food quality, emphasizing proteins and polysaccharides. These materials have unique characteristics that can be used alone or in combination, offering better protection properties and controlled release of probiotics. In addition to

* Corresponding Author's Email: carinilelis@yahoo.com.br.

In: Microencapsulation of Probiotics
Editors: D. de Almeida Paula, K. Cristina Guedes Silva and N. de Almeida Costa
ISBN: 979-8-88697-979-4

seeking new materials for encapsulating probiotics, many research groups aim to improve the materials already in use, varying encapsulation techniques, combinations between various wall materials, and chemical modification.

**Keywords:** protein, polysaccharide, cell viability

## Introduction

Probiotics are live microbial cells known for their health benefits. The health benefits associated with probiotics depend on the administration dose, viability, stability in the applied matrix (e.g., food), survival in the gastrointestinal tract environment, and during storage. Thus, probiotics must pass through the digestive system, support digestive enzymatic hydrolysis, and reach the lower intestinal area and colon for effective absorption. If applied to food, they must remain viable and in recommended doses throughout the product's entire storage and shelf life.

It is recommended that probiotic products contain not less than $10^6$-$10^7$ CFU live cells/mL or CFU live cells/g by the end of life to demonstrate beneficial effects. However, probiotics are extremely sensitive to environmental stresses such as oxygen, lack of nutrients, water activity, temperature, and pH (Šipailienė & Petraitytė, 2018; Z. Zhang et al., 2021).

An alternative to protect probiotic cells is through a physical barrier such as encapsulation that can withstand environmental, processing, storage, and intestinal conditions that are harmful to the survival of these living microorganisms (Arepally & Goswami, 2019). Encapsulation packs substances (probiotics), protecting and releasing them in a controlled manner and under specific conditions. The choice of wall material used in encapsulation is significant and evaluated. These substances must be of food quality, capable of acting as a physical barrier, protecting the core (probiotic) from the external environment, emulsifying properties, film formation, gelling agents, etc., essential depending on the technique used for encapsulation. In Table 1, we present wall materials that act as carriers and protectors for probiotics, guaranteeing the delivery of viable cells to the gastrointestinal tract

**Table 1.** Materials and techniques used for probiotics encapsulation

| Probiotic | Encapsulate material | Technique | Conclusion | References |
|---|---|---|---|---|
| *Lactobacillus casei* 01 | Calcium alginate (ALG);<br>Lactulose + calcium alginate (LAC-ALG);<br>resistant starch + calcium alginate (STA-ALG);<br>Lactosucrose LS40L + calcium alginate (LS40L-ALG);<br>Lactosucrose LS55L + calcium alginate (LS55L-ALG);<br>Gellan gum + xanthan gum (GEL-XNT);<br>κ-carragenaan + locust bean gum (CAR-LBG);<br>Calcium alginate with the coating of Chitosan layer (CHI-ALG);<br>Calcium alginate with the coating of DEAE-Sephadex A50 layer (SDEX coat. ALG) | Hydrogel beads - extrusion | The incorporation of prebiotics increases encapsulation efficiency;<br><br>Incorporating resistant starch provided better protection against low pH and bile salt conditions. | (Ta et al., 2021) |
| *Enterococcus mundtii* SRBG1 | Inulin;<br>Maltodextrin;<br>Inulin + maltodextrin;<br>Inulin + sodium alginate;<br>Inulin + maltodextrin + sodium alginate;<br>Maltodextrin + sodium alginate; | Spray-dryer | Inulin + sodium alginate = further reduction in cell viability;<br>Maltodextrin and Maltodextrin + inulin = less reduction in cell viability;<br>Inulin demonstrated a greater effect in protecting *Enterococos* in acidic conditions | (Sakoui et al., 2022) |
| *Lactobacillus rhamnosus* MF00960, *Pediococcus pentosaceus* MF000967, and *Lactobacillus paracasei* DSM20258 | Camel casein;<br>Camel skin gelatin;<br>Sodium Alginate | Injection-lyophilization | Cell death cycle: sodium alginate > camel casein: camel skin gelatin > camel casein camel casein and camel casein: camel skin gelatin showed better total survivability upon simulated gastrointestinal digestion and thermal tolerance than the sodium alginate | (Devarajan et al., 2022) |

**Table 1.** (Continued)

| Probiotic | Encapsulate material | Technique | Conclusion | References |
|---|---|---|---|---|
| *Lactobacillus paracasei* LC-01 | Kelp nanocellulose + calcium alginate | Gas-liquid microfluidic | Viable cells (gastric juice for 2h, pH = 1.2) = ↑ up to 72% and 30% compared to free cells and encapsulated only with alginate, respectively;<br>The presence of kelp nanocellulose delays the disintegration of microcapsules in the intestinal tract;<br>Kelp nanocellulose + calcium alginate has better storage stability (28 days/4°C) compared to calcium alginate alone | (Wang et al., 2022) |
| *Escherichia coli* Nissle 1917 | Cellulose acetate and Poly (vinyl alcohol) (PVA) | electrospinning | Fibers of poly(vinyl alcohol) and cellulose acetate protected probiotic cells against gastric conditions. However, due to the slow release rate, cellulose acetate was unsuitable. | (Çanga & Dudak, 2021) |
| *Pediococcus pentosaceus* S4–3C | Camel milk proteins and wheat starch | Emulsion - lyophilization | Simulated gastrointestinal digestion conditions: camel milk proteins = ↑ 61.12% viability wheat starch – ↑ 33.27% viability<br>Improved inhibition activity of α-glucosidase, dipeptidyl peptidase IV, and porcine pancreatic amylase<br>Control = free cells;<br>Better maintenance of cell viability when encapsulated with camel milk proteins compared to wheat starch applied to Chami cheese, stored for 9 days at 4°C | (Mudgil et al., 2022) |

| Probiotic | Encapsulate material | Technique | Conclusion | References |
|---|---|---|---|---|
| Lactoferm ABY 6 (*Streptococcus salivarius* ssp. *thermophilus* - 80%, *Lactobacillus acidophilus* -13%, *Bifidobacterium bifidum* - 6%, Lactobacillus delbrueckii ssp. *bulgaricus* -1%. | Sodium alginate - Whey protein Concentrate or Hydrolysis of whey protein concentrate | Extrusion | High encapsulation efficiency > 92%; Improvement in the antioxidant activity and probiotic stability of a whey-based substrate; Hydrolysis positively affected the antioxidant properties and the number of viable cells. | (Krunić & Rakin, 2022) |
| *Pediococcus acidilactici* | Camel whey and cow whey | Emulsion- freeze-dried | ~55% increase in α-glucosidase inhibition after simulated gastrointestinal digestion conditions, with no significant difference in dipeptidyl peptidase IV inhibition compared;<br>Increase between 15-23% and 22-31% in inhibition of pancreatic lipase and cholesterol esterase, respectively, after simulated gastrointestinal digestion conditions; Control = free cells; | (Ahmad et al., 2019) |
| *Limosilactobacillus fermentum* NCU001464 | Gelatin and whey protein isolate-xylooligosaccharides | Spray-drying | ↑ resistance to simulated gastrointestinal digestion; The marked increase in adherence to Caco-2 intestinal epithelial cells after simulated digestion. | (Li et al., 2023) |

**Table 1.** (Continued)

| Probiotic | Encapsulate material | Technique | Conclusion | References |
|---|---|---|---|---|
| *Lactobacillus plantarum* | Alginate-gelatin | Extrusion | ↑ up to 1 log cycle in cell viability in pH 4.5 and 6.5 medium for 120 min, compared to free cells; Alginate-gelatin combination was better at protecting cells in the mimic milk tea system than alginate alone | (Ni et al., 2023) |
| *Lactobacillus acidophilus* | Soy protein isolate and sodium alginate | Double emulsion | Increased viability during gastrointestinal digestion | (Jiang et al., 2023) |
| *Lactobacillus plantarum* 550 | Soy protein isolate and peach gum polysaccharide | Spray drying | Better survival under conditions of gastrointestinal simulation and heat treatment (65°C for 30 min or 75°C for 10 min) | (Yao et al., 2023) |
| *Lactobacillus acidophilus* LA-5 and *Bifidobacterium animalis* BB-12 | High methoxyl pectin | Freeze drying | ↑ cell viability (> 6 Log CFU/mL) compared to free cells after simulated gastrointestinal conditions. | (Motalebi Moghanjougi et al., 2021) |
| *Lactobacillus acidophilus and Lactobacillus plantarum* | Chitosan | Emulsion | ↑ viability of probiotic cells (> 6 Log CFU/mL) during storage of pineapple juice for 20 days/4°C, compared to free cells;<br>Better overall acceptance of pineapple juice containing L. acidophilus and L. plantarum encapsulated with chitosan than pineapple juice containing free probiotic cells. | (Langaroudi et al., 2023) |

| Probiotic | Encapsulate material | Technique | Conclusion | References |
|---|---|---|---|---|
| Mixture: *Lactobacillus delbrueckii* ssp. *bulgaricus*1.0%, *Bifidobacterium bifidum* -6.0%, *Streptococcus salivarius* ssp. *thermophilus*, -80.0%, *Lactobacillus acidophilus* -13.0%. | Whey, whey protein concentrate, and sodium alginate | Freeze and spray-drying | cell survival after fermentation and simulated gastrointestinal conditions compared to free cells | (Obradović et al., 2022) |
| *Lactobacillus acidophilus* La-14 | Sodium alginate and whey protein | extrusion/external ionic gelation | Sodium alginate coat with whey protein layer = ↑ cell viability after simulated gastrointestinal conditions, thermal treatment, and storage compared to free and encapsulated cells only with sodium alginate | (Araújo Etchepare et al., 2020) |
| *Streptococcus thermophilus* NCDC 295 and *Lactobacillus rhamnosus* NCDC 18 | Soy protein isolate | Emulsion-electrospraying and Emulsion- freeze-drying | Simulated Oral Phase:<br>↓ survival of probiotics in both encapsulation processes; Simulated Gastric Phase:<br>Without effect for *S. thermophilus* and L. rhamnosus, by electrospinning and lyophilization, respectively;<br>↓ survival of *S. thermophilus* by lyophilization;<br>↑ survival of *L. rhamnosus* by electrospinning<br>Silumated Intestinal Phase:<br>↑ survival of both probiotics in both encapsulation processes;<br>Control: free cells | (Premjit & Mitra, 2023) |

**Table 1.** (Continued)

| Probiotic | Encapsulate material | Technique | Conclusion | References |
|---|---|---|---|---|
| *Lactobacillus acidophilus* | Sodium alginate, soy protein isolate, and sodium alginate–soy protein isolate | Extrusion | ↑ survival in thermal (50, 63, and 72°C for 2 min), acid (pH 2 and 3), simulated gastric (up to 100 min), simulated intestinal (up to 100 min) conditions, and in the desert during storage (120 days) for both wall materials<br>Control: free cells | (Zeashan et al., 2020) |
| *Lactobacillus acidophilus* | Zein and sodium alginate | Electrospinning | ↑ up to 3 log CFU/mL survival after simulated gastric fluid<br>Control: free cells | (Laelorspoen et al., 2014) |
| *Lactiplantibacillus plantarum* | Maltodextrin-sweet potato-pectin and maltodextrin-sweet potato-gelatin | Spray- drying | Gastric phase: probiotic cell viability = 4 log $CFU.g^{-1}$ higher for encapsulated than free cells;<br>Encapsulated probiotics showed greater survival in the enteric phase than free probiotics | (Costa et al., 2023) |
| *Bifidobacterium bifidum* | Zein and sodium alginate | Extrusion | ↑ survival after 60, 120, and 180 min incubation at pH 2 in simulated gastric fluid, 2% bile salt solution, and storage for 32 days at 4°C<br>Control: free cells and encapsulation with sodium alginate | (Riaz et al., 2019) |

## Carbohydrate Polymers

Alginate is a linear heteropolysaccharide of D-mannuronic and L-guluronic acid extracted from various species of algae. The sequence and composition of L-guluronic acid and D-mannuronic acid vary depending on the specie, and alginate's functional properties depend on this sequence and composition.The main material used to encapsulate probiotics by extrusion is alginate. Alginate is non-toxic, biocompatible, low-cost, harmless, and insoluble in acidic media (Gbassi & Vandamme, 2012; Ni et al., 2023; Skják-Bræk et al., 1986). The beads formation occurs when a cell suspension is mixed with a sodium alginate solution and dripped into a multivalent cation solution. The $Ca^{2+}$ in the form of $CaCl_2$ is the most used (Martinsen et al., 1989; Ni et al., 2023).

Although the use of the extrusion encapsulation method using alginate is considered low-cost, simple, non-toxic, and biocompatible, probiotics may have limitations such as low stability in the presence of anti-gelling cations and $Ca^{2+}$ replacement (e.g., sodium and magnesium ions) and substances with high affinity for calcium such as phosphate and citrate. The high porosity and low pH sensitivity (such as gastric acid) of alginate have been associated with rapid and early leakage of probiotic cells (Ta et al., 2021; Wang et al., 2022). These limitations lead to the need to promote changes in the alginate structure or use other associated carbohydrate polymers (Ni et al., 2023).

In work carried out by Ta et al., (2021), in addition to increasing the efficiency of encapsulation of *Lactobacillus casei* cells when calcium alginate was associated with other polysaccharides, the combination of modified starch with calcium alginate demonstrated greater protection for *Lactobacillus casei* cells after 135 min of gastric treatment and 150 min of simulated intestinal treatment. This greater viability of cells protected by calcium alginate and modified starch can be associated with the structure formed between the two compounds, better protecting the probiotic cells from digestion and treatment with bile salts.

Calcium alginate associated with polysaccharides isolated from Choerospondias axillaries (CP) and emulsion stabilized with CP showed greater antioxidant activity, demonstrating the better potential to improve the resistance of probiotics to oxidation damage. Furthermore, adding CP and CP-coated oil droplets eliminated the negative effect on probiotic viability caused by pure calcium alginate microgels. CP-containing microgels have also been shown to improve the storage and gastrointestinal stability of *Lactobacillus casei* (Li et al., 2023).

Nanocellulose in nanofibers and nanocrystals has emerged recently as a delivery material for probiotics, mainly due to its excellent mechanical properties, biocompatibility, and high surface area (Çanga & Dudak, 2021; Khoshnevisan et al., 2018; Wang et al., 2022). It can interact with hydrophilic polymers, such as alginate, thus reducing its size. Microcapsules prepared with calcium alginate and kelp nanocellulose showed good thermal stability. The probiotics' microcapsules can maintain a complete shape in the stomach to reduce the erosion of gastric acid and disintegrate in the intestinal tract. A better effect on probiotic (*Lactobacillus paracasei)* viability was observed when microcapsules were prepared with the addition of kelp nanocellulose compared to those prepared with alginate alone. The authors suggest that kelp nanocellulose blocks the pores on the alginate surface, preventing ion exchange and gastric acid erosion. After simulated gastric, kelp nanocellulose/alginate microcapsules could maintain viable >7 log CFU/g. Kelp nanocellulose also regulated the release time and rate of probiotics in the intestinal tract, delaying the disintegration time of microcapsules and improving storage stability (28 days/4°C) (Wang et al., 2022).

The acetylation of cellulose obtains cellulose acetate. This material demonstrates high thermal and mechanical stability (Khoshnevisan et al., 2018). Poly (vinyl alcohol) fibers added with cellulose acetate have been shown to reduce cell viability less (*Escherichia coli* Nissle 1917) after simulated gastrointestinal conditions compared to fibers produced only with poly (vinyl alcohol). The authors associate better cell viability with the low release rate of cellulose acetate, especially at low pH, providing better fiber stability for the microorganism. However, none of the fibers produced were effective in cell survival in aqueous medium and dry forms during long-term storage (12 weeks) (Çanga & Dudak, 2021).

Pectin is considered a non-toxic and inexpensive polymer. In the presence of divalent metal ions such as calcium, they form a gel structure. In encapsulation, high molecular weight pectin and high gelling power provide small microparticles (Motalebi Moghanjougi et al., 2021).

Encapsulation of *Lactobacillus acidophilus, Lactobacillus casei,* and *Lactobacillus rhammosus* using low-methoxyl pectins allowed better cell survival upon heat shock (95°C/1min) than free cells. The physical barrier provided by pectin probably retards external heat transfer. Satisfactory protection was also observed during simulated gastric phases. At low pH (pH = 2.0) a denser crosslinking network is formed, retarding gastric acid permeation. In intestinal digestion, at pH = 7.0, electrostatic repulsions between molecular chains allow bile salts to penetrate the particles, coming into contact with *lactobacilli*, reducing their viability. For freeze-drying, the same protection was not observed (He et al., 2021). High methoxyl pectin used for encapsulating *Lactobacillus acidophilus* and *Bifidobacterium animalis* was also able to protect probiotic cells against simulated gastrointestinal conditions, maintaining viability above 6 Log CFU/mL (Motalebi Moghanjougi et al., 2021).

Maltodextrin is derived from corn starch's enzymatic and/or acid hydrolysis, which contains linear amylose and branched amylopectin degradation products (Rezende & Hashizume, 2018; Wangsakan et al., 2001). Maltodextrin is a non-toxic, low-cost, high-availability wall material with physical and thermal resistance, good solubility, and low viscosity (Anekella & Orsat, 2013). However, especially in low-viscosity products, their use alone can precipitate the release of encapsulated material, such as probiotic cells (Francisco et al., 2018).

The presence of maltodextrin also positively affects the survival of microencapsulated *Enterococci*. This is because, during drying (spray drying), maltodextrin probably has a thermoprotective effect and thus prevents the rupture of cell membranes (Sakoui et al., 2022).

Wheat starch features 20%-30% amylose and 70%-80% amylopectin. It is used as a thickener, stabilizer, and fat substitute (Ge et al., 2023; Xie et al., 2022; C. Zhang et al., 2023; Zheng et al., 2023). Second (Mudgil et al., 2022) wheat starch demonstrated a 33.27% increase in the viability of *Pediococcus pentosaceus* probiotic cells after simulated gastrointestinal digestion compared to free cells. To provide the host's health benefits, a minimum of $10^6$ -$10^7$ CFU/mL or grams of food is required. The viability of Pediococcus pentosaceus encapsulated with wheat starch when applied to Chami cheese showed 6.56 Log CFU/g after 9 days of storage at 4C. Improved inhibitory properties of α-glucosidase, dipeptidyl peptidase IV and porcine pancreatic amylase enzymes were also observed in Chami cheese compared to the addition of free cells.

Chitosan is a poly-cationic linear polysaccharide composed of randomly distributed β-linked D-glucosamine and N-acetyl-D-glucosamine. The intrinsic characteristics of chitosan may allow it to generate electrostatic complexes or multilayer structures with other negatively-charged polymers (Cheung et al., 2015; Langaroudi et al., 2023). The solubility of chitosan is pH dependent, being insoluble in water at pH > 5.4 (Gbassi & Vandamme, 2012).

Chitosan used alone and associated with Tragacanth gum has been shown to maintain better viability of probiotic cells (>6 Log CFU/mL) *Lactobacillus acidophilus* and *Lactobacillus plantarum*, during storage of pineapple juice for 20 days/4°C, compared to free cells (<6 Log CFU/mL). Satisfactorily, the pineapple juice samples containing *L. acidophilus* and *L. plantarum* encapsulated with chitosan showed better global acceptance compared to cells applied free (Langaroudi et al., 2023).

## Protein

Casein is the major protein in milk in the form of colloidal micelles, formed by the fractions $\alpha s_1$-, $\alpha s_2$-, β- and k-casein. They are known as a carrier for calcium, phosphorus, and other micronutrients. Either in their micellar form, individual caseins or casein salts (calcium and sodium caseinate) are used as emulsifiers and wall materials to encapsulate various compounds. On the other hand, whey proteins represent about 20% of the total proteins in milk, with high nutritional value and emulsification/foaming and gelling properties. Camel casein demonstrated a shorter cycle of death for *Lactobacillus rhamnosus* MF00960,

*Pediococcus pentosaceus* MF000967, and *Lactobacillus paracasei* DSM20258 compared to the use of sodium alginate. According to the authors, milk proteins present better gelling properties and high viscosity, which provides better dispersion and greater protection to encapsulated microorganisms. The buffering capacity of proteins also helps protect against aggressive environments. Functional and bioactive properties aid cell viability (Devarajan et al., 2022). Second Mudgil et al. (2022) camel milk proteins demonstrated a 61.12% increase in the viability of *Pediococcus pentosaceus* probiotic cells after simulated gastrointestinal digestion compared to free cells. The high viability of the probiotic cells (98.59%) encapsulated with camel milk proteins simulated gastrointestinal digestion was associated with the formation of dense and continuous matrices that promoted the total entrapment of the probiotics, isolating them from harmful external conditions. Maintaining probiotic amounts greater than 6 Log $CFUg^{-1}$ suggests that camel milk proteins provide sufficient protection against environmental stresses, maintaining survival after exposure to gastrointestinal fluids.

When applied to Chami cheese, only a 0.41 Log CFU $g^{-1}$ decline in probiotic cells was observed after 9 days of storage at 4°C. Encapsulation also improved the inhibition properties of α-glucosidase, dipeptidyl peptidase IV and porcine pancreatic amylase (diabetes control) in Chami cheese. This demonstrates that improving the viability of probiotic cells, greater activity, and stability promoted by encapsulation using camel milk proteins can improve the therapeutic benefits of Chami cheese (Mudgil et al., 2022). After simulated gastrointestinal digestion conditions, an increase of approximately 55% in α-glucosidase inhibition was also observed for *Pediococcus acidilactici* encapsulated with camel whey and cow whey. However, there was no significant difference compared to free cells for inhibiting dipeptidyl peptidase IV. In addition to the enzymes associated with diabetes problems mentioned above, the inhibition of metabolic enzymes associated with obesity, such as pancreatic lipase and cholesterol esterase, has also been documented for *Pediococcus acidilactici* encapsulated with camel whey and cow whey, subjected to simulated gastrointestinal digestion conditions. The encapsulation process with whey proteins showed a significant protective effect in retaining the inhibitory potential against these enzymes. Whey protein showed an improved protective effect compared to camel whey protein (Ahmad et al., 2019).

Protein hydrolysis processes have been studied to improve the application of milk proteins in the encapsulation of probiotics. Whey protein hydrolysates have been shown to promote better maintenance of a mixture of probiotic cells applied on a whey-based substrate during 28 days of storage at 4°C. Higher antioxidant activity was also observed (Krunić & Rakin, 2022).

Gelatin, a protein, derived from denatured collagen, contains high hydroxyproline, glycine, and proline levels. It presents hydroxyl and hydrophobic groups in its structure, a polyampholyte with positive and negative charges. It acts as a thermally reversible gelling agent and has excellent membrane formation capacity, biocompatibility, and non-toxicity. However, its low network rigidity limits its applicability as a hydrogel matrix. With that, crosslinking agents have been added to improve their physical properties. They are amphoteric, cooperating with anionic polysaccharides such as alginate (Gbassi & Vandamme, 2012; X. Y. Li et al., 2009; Ni et al., 2023; Sun & Griffiths, 2000).

Camel skin gelatin showed better encapsulation properties for *Lactobacillus rhamnosus*, *Pediococcus pentosaceus*, and *Lactobacillus paracasei* than sodium alginate (Devarajan et al., 2022).

The association of gelatin with whey protein isolate-xylooligosaccharides (XOS), an oligosaccharide, has been proposed to improve gelatin's properties as a wall material. The XOS act as prebiotics, helping in the proliferation of bifidobacteria. They help promote health and are usually stable in acidic pH and high temperatures, as well as in the hydrolysis of enzymes in the gastrointestinal tract. These characteristics make the association of gelatin and XOS by Maillard reaction proposed as a material to encapsulate

*Limosilactobacillus fermentum*. After spray drying, the viable *Limosilactobacillus fermentum* count was 3 x $10^9$ – 6 x $10^9$ CFU/g. XOS obtained by the Maillard reaction promoted resistance of the cells to simulated gastrointestinal digestion and adhesion to intestinal epithelial cells (Caco-2), after simulated digestion, demonstrating the ability to colonize the colon (Li et al., 2023).

Soy proteins have been extensively used in the food and pharmaceutical industry due to their functional properties, low cost, and high nutritional value (Yew et al., 2011). Despite the high viability of free cells in the oral phase, probiotics must present sufficient amounts after conditions in the gastrointestinal tract and reach the colon to confer health benefits. It was found that encapsulation by electrospinning using soy protein showed no significant difference in the simulated gastric phase for *Streptococcus thermophilus*, but demonstrated better viability (an increase of 12%) for *Lactobacillus rhamnosus* compared to free cells. In the simulated intestinal phase, encapsulation by electrospinning showed greater viability for both microorganisms than free cells. Better results were achieved using the electrospinning process to encapsulate these microorganisms than the lyophilization process (Premjit & Mitra, 2023). Soy protein isolate has been shown to protect *Lactobacillus acidophilus* from conditions such as heat treatment, acidic pH, simulated gastric and intestinal, and when added in dessert compared to free cells (Zeashan et al., 2020).

## Polysaccharides-Protein

The combination of different wall materials can be strategic for the encapsulation of probiotics. By uniting the other properties of each material or even forming new properties after combining them, significant improvements can be observed in systems for probiotics encapsulated in these matrices. The use of proteins and polysaccharides as encapsulating materials is associated mainly with their interactions. The positive charge of proteins when the pH is adjusted below the isoelectric point facilitates (electrostatic) interactions with negatively charged polysaccharides (Gbassi & Vandamme, 2012).

The combination of alginate and gelatin has been shown to increase the viability of *Lactobacillus plantarum* by more than 1 log cycle after immersion in solution, both at pH 4.5/120 min and pH 6.5/120 min, compared to free cells. The alginate and gelatin combination demonstrated better protection of

*Lactobacillus plantarum* when in direct contact with the medium in the mimic milk tea system than encapsulation only with alginate (Ni et al., 2023). During the experiment, thermal resistance, *in vitro* release behavior, storage stability, and physical-chemical properties were evaluated. According to the authors, this encapsulation type promotes the protection and distribution of probiotic microorganisms. Thus, the combination of these wall materials has great potential for application in various products in the food industry.

Costa et al., (2023) studied, by spray drying, the encapsulation of *Lactiplantibacillus plantarum* using the combination of maltodextrin-sweet potato-pectin and maltodextrin-sweet potato-gelatin. In the gastric phase, the reduced viability of encapsulated probiotics was approximately 4 log CFU.$g^{-1}$ lower than free cells. Highlight in the survival of encapsulated probiotic cells compared to free cells was also observed after the simulated enteric phase.

When applied in whey-based beverage, it was verified that probiotic cells (*Lactobacillus delbrueckii* ssp. *Bulgaricus*, *Bifidobacterium bifidum*, *Streptococcus salivarius* ssp. *Thermophilus*, and *Lactobacillus acidophilus*) encapsulated with whey, whey protein concentrate, and sodium alginate showed better results for viability after fermentation. Moreover, encapsulation has been shown to extend fermentation time compared to free cells. Compared to free cells, encapsulated cells showed better viability after storage for 28 days at 25°C. In addition, it was found that encapsulation using the spray-drying technique provided better viability, explained by structural changes in the protein caused by heat. In simulated gastrointestinal conditions, it was found that the survival of free cells was 41.86% and 27.97% after 2h and 4h, respectively. For encapsulated cells, the survival of free cells was >83% after 4 hours (Obradović et al., 2022). Improved viability of *Lactobacillus acidophilus* after simulated gastrointestinal conditions, thermal treatment (72°C for 15s and 5 min and 63°C for 30, 60, and 120 min), and storage conditions (7°C, -18°C, and 25°C for up to 120 days) was verified when coating the sodium alginate particles containing the probiotic cells with whey protein (Araújo Etchepare et al., 2020).

*Lactobacillus rhamnosus*, *Pediococcus pentosaceus*, and *Lactobacillus paracasei* remained more resistant to the harmful effects of hydrogen ions and pepsin from gastric fluids when encapsulated with camel casein and camel casein: camel skin gelatin compared to sodium alginate. Encapsulation of probiotics using camel casein and camel casein: camel skin gelatin was also shown to be more efficient in thermal tolerance than cells encapsulated only with sodium alginate or camel skin gelatin alone. These findings were correlated with the structural integrity and rigidity of these matrices as an encapsulation material for probiotics, maintaining their viability during adverse environmental conditions (Devarajan et al., 2022).

Encapsulation of *Lactobacillus plantarum* using soy protein isolate and peach gum polysaccharide allowed cell survival rates to be maintained above 6 log CFU/g during simulated gastric (1-2h) and intestinal (2-6h) digestions, unlike free cells, which lost all their viability in just 0.5h. Adding peach gum polysaccharide is critical in maintaining cell survival after simulated gastrointestinal digestion. The electrostatic interaction between soy protein (positively charged) and peach gum polysaccharide (negatively charged) under gastric conditions led to more compact structures forming extra protection for the encapsulated probiotic cells. The ratio of soy protein: peach gum polysaccharide = 3:1 retained the maximum activity of 7.88 Log CFU/g. Under heat treatment conditions (65°C for 30 min or 75°C for 10 min), the encapsulation process also improved the viability of probiotic cells compared to heat-treated free cells (Yao et al., 2023). Soy protein isolate associated with sodium alginate showed potential to protect *Lactobacillus acidophilus* subjected to heat treatment, acidic pH, simulated gastric and intestinal, and when added in the dessert. The sodium alginate-soy protein isolate combination showed better results compared to wall materials used alone (Zeashan et al., 2020). After passing through the simulated gastrointestinal, it was verified that the structure W1/O/W2 prepared with soy protein, and fish oil added with sodium alginate showed greater viability of *Lactobacillus acidophilus* (Jiang et al., 2023).

Zein, the main protein in corn, has in recent years been shown to be an interesting wall material for encapsulation due to its resistance to gastric juice and amphiphilic characteristics (Riaz et al., 2019). Zein and sodium alginate were combined to encapsulate *Lactobacillus acidophilus* by electrospinning. After simulated gastric fluid (pH 1.2/2 hours), only 2 log CFU/mL decreased, compared to ~5 log CFU/mL decrease for free cells (Laelorspoen et al., 2014). Higher viability of *Bifidobacterium bifidum* after 60, 120, and 180 min incubation at pH2 in simulated gastric fluid, 2% bile salt solution, and storage for 32 days at 4°C was verified after encapsulation using zein and sodium alginate, compared to cells free and the use of sodium alginate alone (Riaz et al., 2019).

## Conclusion

Probiotics have been associated with numerous benefits to human health and well-being. However, they must be ingested sufficiently to promote their beneficial effects. Their viability must be maintained after adverse conditions related to food composition and body conditions, mainly of the gastrointestinal tract. In general, encapsulation can be used to protect probiotics from these adverse conditions, needing to be carefully designed to ensure the viability and bioactivities of the probiotics. A variety of encapsulation systems are available for the application of probiotics. This chapter addresses the main materials used for encapsulation, regardless of the technique used. As we can see, many food ingredients (e.g., proteins and polysaccharides) have demonstrated potential as encapsulation matrices for probiotics. Many studies have used these materials to encapsulate probiotics, improving their properties as a wall material. The search for new materials for encapsulation grows every year, and many research groups work extensively to improve the properties of materials already proposed.

## References

Ahmad, M., Mudgil, P., & Maqsood, S. (2019). Camel whey protein microparticles for safe and efficient delivery of novel camel milk-derived probiotics. *LWT*, *108*, 81–88. https://doi.org/10.1016/j.lwt.2019.03.008

Anekella, K., & Orsat, V. (2013). Optimization of microencapsulation of probiotics in raspberry juice by spray drying. *LWT - Food Science and Technology*, *50*(1), 17–24. https://doi.org/10.1016/j.lwt.2012.08.003

Araújo Etchepare, M., Nunes, G. L., Nicoloso, B. R., Barin, J. S., Moraes Flores, E. M., de Oliveira Mello, R., & Ragagnin de Menezes, C. (2020). Improvement of the viability of encapsulated probiotics using whey proteins. *LWT*, *117*, 108601. https://doi.org/10.1016/j.lwt.2019.108601

Arepally, D., & Goswami, T. K. (2019). Effect of inlet air temperature and gum Arabic concentration on encapsulation of probiotics by spray drying. *LWT*, *99*, 583–593. https://doi.org/10.1016/j.lwt.2018.10.022

Çanga, E. M., & Dudak, F. C. (2021). Improved digestive stability of probiotics encapsulated within poly(vinyl alcohol)/cellulose acetate hybrid fibers. *Carbohydrate Polymers*, *264*, 117990. https://doi.org/10.1016/j.carbpol.2021.117990

Cheung, R., Ng, T., Wong, J., & Chan, W. (2015). Chitosan: An Update on Potential Biomedical and Pharmaceutical Applications. *Marine Drugs*, *13*(8), 5156–5186. https://doi.org/10.3390/md13085156

Costa, N. de A., Silveira, L. R., Amaral, E. de P., Pereira, G. C., Paula, D. de A., Vieira, É. N. R., Martins, E. M. F., Stringheta, P. C., Leite Júnior, B. R. de C., & Ramos, A. M. (2023). Use of maltodextrin, sweet potato flour, pectin, and gelatin as wall material for microencapsulating Lactiplantibacillus plantarum by spray drying: Thermal resistance, *in vitro* release behavior, storage stability, and physicochemical properties. *Food Research International*, *164*, 112367. https://doi.org/10.1016/j.foodres.2022.112367

Devarajan, A., Mudgil, P., Aldhaheri, F., Hamed, F., Dhital, S., & Maqsood, S. (2022). Camel milk-derived probiotic strains encapsulated in camel casein and gelatin complex microcapsules: Stability against thermal challenge and simulated gastrointestinal digestion conditions. *Journal of Dairy Science*, *105*(3), 1862–1877. https://doi.org/10.3168/jds.2021-20745

Francisco, C. R. L., Heleno, S. A., Fernandes, I. P. M., Barreira, J. C. M., Calhelha, R. C., Barros, L., Gonçalves, O. H., Ferreira, I. C. F. R., & Barreiro, M. F. (2018). Functionalization of yogurts with Agaricus bisporus extracts encapsulated in spray-dried maltodextrin crosslinked with citric acid. *Food Chemistry*, *245*, 845–853. https://doi.org/10.1016/j.foodchem.2017.11.098

Gbassi, G. K., & Vandamme, T. (2012). Probiotic Encapsulation Technology: From Microencapsulation to Release into the Gut. *Pharmaceutics*, *4*(1), 149–163. https://doi.org/10.3390/pharmaceutics4010149

Ge, Z., Gao, S., Xu, M., Zhao, Y., Wei, X., Zong, W., & Zhao, G. (2023). Effects of Deacetylated Konjac Glucomannan on the pasting, rheological, and retrogradation properties of wheat starch. *Food Bioscience*, *52*, 102465. https://doi.org/10.1016/j.fbio.2023.102465

He, C., Sampers, I., Van de Walle, D., Dewettinck, K., & Raes, K. (2021). Encapsulation of Lactobacillus in Low-Methoxyl Pectin-Based Microcapsules Stimulates Biofilm Formation: Enhanced Resistances to Heat Shock and Simulated Gastrointestinal Digestion. *Journal of Agricultural and Food Chemistry*, *69*(22), 6281–6290. https://doi.org/10.1021/acs.jafc.1c00719

Jiang, Z., Tian, J., Bai, X., McClements, D. J., Ma, C., Liu, X., & Liu, F. (2023). Improving probiotic survival using water-in-oil-in-water (W1/O/W2) emulsions: Role of fish oil in inner phase and sodium alginate in outer phase. *Food Chemistry*, *417*, 135889. https://doi.org/10.1016/j.foodchem.2023.135889

Khoshnevisan, K., Maleki, H., Samadian, H., Shahsavari, S., Sarrafzadeh, M. H., Larijani, B., Dorkoosh, F. A., Haghpanah, V., & Khorramizadeh, M. R. (2018). Cellulose acetate electrospun nanofibers for drug delivery systems: Applications and recent advances. *Carbohydrate Polymers*, *198*, 131–141. https://doi.org/10.1016/j.carbpol.2018.06.072

Krunić, T. Ž., & Rakin, M. B. (2022). Enriching alginate matrix used for probiotic encapsulation with whey protein concentrate or its trypsin-derived hydrolysate: Impact on antioxidant capacity and stability of fermented whey-based beverages. *Food Chemistry*, *370*, 130931. https://doi.org/10.1016/j.foodchem.2021.130931

Laelorspoen, N., Wongsasulak, S., Yoovidhya, T., & Devahastin, S. (2014). Microencapsulation of Lactobacillus acidophilus in zein–alginatecore–shell microcapsules via electrospraying. *Journal of Functional Foods*, *7*, 342–349. https://doi.org/10.1016/j.jff.2014.01.026

Langaroudi, S. S., Nouri, L., & Azizi, M. H. (2023). Influence of encapsulation with chitosan and tragacanth gum on the physicochemical and overall impression of probiotic pineapple juice. *Journal of Food Measurement and Characterization*, *17*(2), 1382–1392. https://doi.org/10.1007/s11694-022-01712-8

Li, H., Peng, F., Lin, J. X., Xiong, T., & Huang, T. (2023). Preparation of probiotic microcapsules using gelatin-xylooligosaccharides conjugates by spray drying: Physicochemical properties, survival, digestion resistance, and colonization. *Food Bioscience*, *52*, 102462. https://doi.org/10.1016/j.fbio.2023.102462

Li, Q., Lin, H., Li, J., Liu, L., Huang, J., Cao, Y., Zhao, T., McClements, D. J., Chen, J., Liu, C., Liu, J., Shen, P., & Zhou, M. (2023). Improving probiotic (Lactobacillus casei) viability by encapsulation in alginate-based microgels: Impact of polymeric and colloidal fillers. *Food Hydrocolloids*, *134*, 108028. https://doi.org/10.1016/j.foodhyd.2022.108028

Li, X. Y., Chen, X. G., Cha, D. S., Park, H. J., & Liu, C. S. (2009). Microencapsulation of a probiotic bacteria with alginate–gelatin and its properties. *Journal of Microencapsulation*, *26*(4), 315–324. https://doi.org/10.1080/02652040802328685

Martinsen, A., Skjåk-Braek, G., & Smidsrød, O. (1989). Alginate as immobilization material: I. Correlation between chemical and physical properties of alginate gel beads. *Biotechnology and Bioengineering*, *33*(1), 79–89. https://doi.org/10.1002/bit.260330111

Motalebi Moghanjougi, Z., Rezazadeh Bari, M., Alizadeh Khaledabad, M., Amiri, S., & Almasi, H. (2021). Microencapsulation of Lactobacillus acidophilus LA-5 and Bifidobacterium animalis BB-12 in pectin and sodium alginate: A comparative study on viability, stability, and structure. *Food Science & Nutrition*, *9*(9), 5103–5111. https://doi.org/10.1002/fsn3.2470

Mudgil, P., Aldhaheri, F., Hamdi, M., Punia, S., & Maqsood, S. (2022). Fortification of Chami (traditional soft cheese) with probiotic-loaded protein and starch microparticles: Characterization, bioactive properties, and storage stability. *LWT*, *158*, 113036. https://doi.org/10.1016/j.lwt.2021.113036

Ni, F., Luo, X., Zhao, Z., Yuan, J., Song, Y., Liu, C., Huang, M., Dong, L., Xie, H., Cai, L., Ren, G., & Gu, Q. (2023). Enhancing viability of Lactobacillus plantarum encapsulated by alginate-gelatin hydrogel beads during gastrointestinal digestion, storage, and in the mimic beverage systems. *International Journal of Biological Macromolecules*, *224*, 94–104. https://doi.org/10.1016/j.ijbiomac.2022.10.106

Obradović, N., Volić, M., Nedović, V., Rakin, M., & Bugarski, B. (2022). Microencapsulation of probiotic starter culture in protein: carbohydrate carriers using spray and freeze-drying processes: Implementation in whey-based beverages. *Journal of Food Engineering*, *321*, 110948. https://doi.org/10.1016/j.jfoodeng.2022.110948

Premjit, Y., & Mitra, J. (2023). Synthesis, characterization, and *in vitro* digestion of electrospray and freeze-dried probiotics encapsulated in soy protein isolate-sunflower oil emulsions. *Food Bioscience*, *53*, 102532. https://doi.org/10.1016/j.fbio.2023.102532

Rezende, G., & Hashizume, L. N. (2018). Maltodextrin and dental caries: a literature review. *RGO - Revista Gaúcha de Odontologia*, *66*(3), 257–262. https://doi.org/10.1590/1981-863720180003000103288

Riaz, T., Iqbal, M. W., Saeed, M., Yasmin, I., Hassanin, H. A. M., Mahmood, S., & Rehman, A. (2019). *In vitro*, survival of Bifidobacterium bifidum microencapsulated in zein-coated alginate hydrogel microbeads. *Journal of Microencapsulation*, *36*(2), 192–203. https://doi.org/10.1080/02652048.2019.1618403

Sakoui, S., Derdak, R., Pop, O. L., Vodnar, D. C., Addoum, B., Teleky, B.-E., Elemer, S., Elmakssoudi, A., Suharoschi, R., Soukri, A., & El Khalfi, B. (2022). Effect of encapsulated probiotic in Inulin-Maltodextrin-Sodium alginate matrix on the viability of Enterococcus mundtii SRBG1 and the rheological parameters of fermented milk. *Current Research in Food Science*, *5*, 1713–1719. https://doi.org/10.1016/j.crfs.2022.09.027

Šipailienė, A., & Petraitytė, S. (2018). Encapsulation of Probiotics: Proper Selection of the Probiotic Strain and the Influence of Encapsulation Technology and Materials on the Viability of Encapsulated Microorganisms. *Probiotics and Antimicrobial Proteins*, *10*(1), 1–10. https://doi.org/10.1007/s12602-017-9347-x

Skjåk-Bræk, G., Smidsrød, O., & Larsen, B. (1986). Tailoring of alginates by enzymatic modification *in vitro*. *International Journal of Biological Macromolecules*, *8*(6), 330–336. https://doi.org/10.1016/0141-8130(86)90051-6

Sun, W., & Griffiths, M. W. (2000). Survival of bifidobacteria in yogurt and simulated gastric juice following immobilization in gellan–xanthan beads. *International Journal of Food Microbiology*, *61*(1), 17–25. https://doi.org/10.1016/S01681605(00)00327-5

Ta, L. P., Bujna, E., Antal, O., Ladányi, M., Juhász, R., Szécsi, A., Kun, S., Sudheer, S., Gupta, V. K., & Nguyen, Q. D. (2021). Effects of various polysaccharides (alginate, carrageenan, gums, chitosan) and their combination with prebiotic saccharides (resistant starch, lactosucrose, lactulose) on the encapsulation of probiotic bacteria Lactobacillus casei 01 strain. *International Journal of Biological Macromolecules*, *183*, 1136–1144. https://doi.org/10.1016/j.ijbiomac.2021.04.170

Wang, K., Ni, J., Li, H., Tian, X., Tan, M., & Su, W. (2022). Survivability of probiotics encapsulated in kelp nanocellulose/alginate microcapsules on the microfluidic device. *Food Research International*, *160*, 111723.https://doi.org/10.1016/j.foodres.2022.111723

Wangsakan, A., Chinachoti, P., & McClements, D. J. (2001). Maltodextrin−Anionic Surfactant Interactions: Isothermal Titration Calorimetry and Surface Tension Study. *Journal of Agricultural and Food Chemistry*, *49*(10), 5039–5045. https://doi.org/10.1021/jf0103471

Xie, H., Ying, R., & Huang, M. (2022). Effect of arabinoxylans with different molecular weights on the gelling properties of wheat starch. *International Journal of Biological Macromolecules*, *209*, 1676–1684. https://doi.org/10.1016/j.ijbiomac.2022.04.104

Yao, H., Liu, B., He, L., Hu, J., & Liu, H. (2023). The incorporation of peach gum polysaccharide into soy protein-based microparticles improves probiotic bacterial survival during simulated gastrointestinal digestion and storage. *Food Chemistry*, *413*, 135596. https://doi.org/10.1016/j.foodchem.2023.135596

Yew, S. E., Lim, T. J., Lew, L. C., Bhat, R., Mat-Easa, A., & Liong, M. T. (2011). Development of a Probiotic Delivery System from Agrowastes, Soy Protein Isolate, and Microbial Transglutaminase. *Journal of Food Science*, *76*(3), H108–H115. https://doi.org/10.1111/j.1750-3841.2011.02107.x

Zeashan, M., Afzaal, M., Saeed, F., Ahmed, A., Tufail, T., Ahmed, A., & Anjum, F. M. (2020). Survival and behavior of free and encapsulated probiotic bacteria under simulated human gastrointestinal and technological conditions. *Food Science & Nutrition*, *8*(5), 2419–2426. hhttps://doi.org/10.1002/fsn3.1531

Zhang, C., Wang, M., Tan, Z., Ma, M., Sui, Z., & Corke, H. (2023). The differential distribution of surface proteins/lipids between wheat A- and B-starch granule contributes to their difference in pasting and rheological properties. *International Journal of Biological Macromolecules*, 124430. https://doi.org/10.1016/j.ijbiomac.2023.124430

Zhang, Z., Gu, M., You, X., Sela, D. A., Xiao, H., & McClements, D. J. (2021). Encapsulation of bifidobacterium in alginate microgels improves viability and targeted gut release. *Food Hydrocolloids*, *116*, 106634. https://doi.org/10.1016/j.foodhyd.2021.106634

Zheng, L., Zhang, Q., Yu, X., Luo, X., & Jiang, H. (2023). Effect of annealing and heat-moisture pretreatment on the quality of 3D-printed wheat starch gels. *Innovative Food Science & Emerging Technologies*, *84*, 103274. https://doi.org/10.1016/j.ifset.2023.103274

# Chapter 4

# An Ionic Gelation Technique for Microencapsulation

**Karen Cristina Guedes Silva[1,*], PhD and Gabriela Feltre[2], PhD**

[1]Federal Institute of Paraná – IFPR - Campus Assis Chateaubriand, Assis Chateaubriand, Paraná, Brazil

[2]Department of Agri-food Industry, Food, and Nutrition, Luiz de Queiroz College of Agriculture, University of São Paulo - USP, Piracicaba, São Paulo, Brazil

## Abstract

Functional food has been increasingly demanded by people due to its benefits to health and to its capacity to prevent some diseases that are related to diet. Probiotics are known as microorganisms related to beneficial effects, promoting health and wellness for hosts. However, to promote health benefits, probiotics must reach the intestinal tract in sufficient amount, which does not always occur, since they often lose their viability during food processing, transport, or storage. In addition, after consumption, microorganisms are exposed to adverse pH conditions during their passage through the gastric tract. An alternative to avoid or reduce the loss of probiotics viability is microencapsulation. Among the various existing methods, ionic gelation is a promising technique, since it can improve the resistance of probiotics against adverse conditions. Ionic gelation presents some advantages, such as simplicity and low cost, in addition to mild process conditions, which guarantee the viability of probiotics during processing. Microencapsulation by ionic gelation consists of trapping a compound within a three-dimensional network formed by the crosslinking between polysaccharide chains and

* Corresponding Author's Email: karen.silva@ifpr.edu.br.

In: Microencapsulation of Probiotics
Editors: D. de Almeida Paula, K. Cristina Guedes Silva and N. de Almeida Costa
ISBN: 979-8-88697-979-4

oppositely charged ions. There are two different methods of ionic gelation - internal and external – but both have the same principle of binding between the wall material chains and the ions. In addition to the polysaccharide responsible for generating the network, the particles can be formed with the addition of fillers, which allows greater retention of the encapsulated compound. Methodologies such as extrusion, atomization, or emulsification are usually used in combination with ionic gelation for particle production. Several studies used these techniques to microencapsulate probiotics and evaluated the behavior of the particles about the stability during storage and release in the gastrointestinal tract. Microencapsulated probiotics have been added to food products to provide their protection and release in a viable form in the intestinal tract, promoting the desired benefits. The variety of foods added with probiotic particles that have already been studied is wide, and we can highlight dairy products, such as cheese, cream cheese, butter and yogurt, meat products, as well as beverages, vegetables, desserts, and chocolates. In addition to evaluating particles in the gastrointestinal environment, the study of the interaction of particles with the food matrix is extremely important to determine their influence on the physical, nutritional, and sensory properties of the foods to which they are applied.

**Keywords:** probiotics, microencapsulation, electrostatic interactions, food application

## Introduction

Microencapsulation can be performed using a variety of techniques. The application is a determining factor in choosing the process, wall materials, particle size, and release mechanism, as well as the biological and physicochemical properties of both the bioactive and the encapsulating agent (Lu et al. 2021).

Among the various techniques that can be used to encapsulate probiotics, ionic gelation is an interesting alternative, as it is simple, low-cost, and scalable, and the process conditions are mild, which ensures probiotic viability. For the production of encapsulation systems by ionic gelation, non-toxic, edible, and biodegradable biopolymers are applied, such as pectin, carrageenan, and gellan gum, highlighting alginate (Chun, Kim, and Cho 2014).

The ionic gelation process can be performed by two techniques: internal or external gelation. In the internal gelation process, the crosslinking agent is inactive in the biopolymer solution, being ionized through a trigger, such as pH reduction, which promotes gelation. In the external gelation process, the microgels formation occurs through the diffusion of divalent cations across the surface of the formed biopolymer droplet. In general, the internal gelling process provides greater control over crosslinking agent, resulting in the formation of a homogeneous matrix. Alternatively, particles formed by external gelation show a gradient concentration due to the diffusion process (Puguan, Yu, and Kim 2014).

Biopolymer matrices produced by electrostatic interactions have presented positive results, such as great aqueous solubility, resistance to the action of enzymes in the upper part of the gastrointestinal tract, good barrier properties and mechanical resistance, proving to be a good vehicle for bioactive compounds, which enables the protection of probiotics for food applications (Chai et al. 2018; Silva et al. 2018).

Currently, healthy foods have shown a growing demand from consumers, which has driven the development of food products containing probiotics. Several food products can be mentioned, such as dairy products, like cheese, cream cheese, yogurt, and butter; non-dairy products, such as juice, tea, vegetable drinks, and chocolate; meat products, food supplements, and desserts (Bucheli, Fugaban, and Todorov 2022; Rodrigues et al. 2020; Pimentel et al. 2021). These food varieties can benefit consumers with different dietary restrictions, tastes, and habits.

In this chapter, we will discuss the principle of the ionic gelation technique, its advantages, and limitations for the protection of bioactive compounds, with emphasis on probiotic protection. Some techniques associated with ionic gelation will be addressed, which allow us to overcome some obstacles, such as the porosity of alginate particles, one of the most used biopolymers. Technological applications in various foods will be presented through recent research, which evaluated the viability of probiotics encapsulated by ionic gelation in the face of adverse processing, storage, and consumption conditions.

## Ionic Gelation Technique

Ionic gelation is a microencapsulation technique that occurs through crosslinking, that is the bonding between polysaccharide chains and ions of opposite charges. The binding results in a three-dimensional network, which can trap compounds, making it an excellent matrix for microencapsulation

(Burey et al. 2008; Arriola et al. 2019). The network formation occurs when ions, such as calcium or potassium, diffuse into the polysaccharide chain, resulting in a rearrangement and producing a solid matrix (Funami et al. 2009).

The most common polysaccharide used for ionic gelation encapsulation is sodium alginate. Alginate gel is formed by a mechanism of interaction of G segments (guluronic acid) of alginate with cations (e.g., calcium and potassium). This mechanism is explained by the "egg box" model. The alginate G blocks line up, forming a cavity, where the divalent calcium cations bond, forming a three-dimensional network. The structure formed is similar to an egg box, in which the eggs represent the $Ca^{2+}$ cations (Grant et al. 1973). Alginate gels have thermal stability, which allows them to be subjected to thermal treatments (including cooking and sterilization) without altering their shape and rheological characteristics. This characteristic of alginate gels also allows them to be stored at room temperature (Onsoyen 1997). Another polysaccharide widely used in gel formation by ionic gelation is pectin with a low degree of methoxylation (Dickinson 2003).

Ionic gelation can occur through two different mechanisms: external or internal. External ionic gelation occurs when a biopolymer solution is introduced into an ionic solution, at neutral pH. External gelation occurs heterogeneously, as the ions first come into contact with the surface of the biopolymer, reacting instantly. Subsequently, the ions diffuse along the interior of the particle, generating the three-dimensional network (Helgerud et al. 2010; Smrdel et al. 2008; Burey et al. 2008). This technique is very common to encapsulate different compounds, including probiotics, such as *Lactobacillus acidophilus* La-05 (Lopes et al. 2021), *Lactobacillus acidophilus* La-3 (Lopes Neto et al. 2021), *Lactobacillus acidophilus* and *Bifidobacterium bifidum* (Naissinger da Silva, Tagliapietra, and dos Santos Richards 2021).

The principle of internal ionic gelling is the release of calcium ions into the solution, which is complex with the carboxylic chains of the polymer. The mechanism happens as follows: a salt is added to the polysaccharide solution and there is the addition of an acid, which makes the salt soluble and releases calcium ions, generating the gel (Poncelet et al. 1992; Naranjo-Durán et al. 2021; Lupo et al. 2015; Funami et al. 2009). The most commonly used salt is calcium carbonate. Some studies encapsulated compounds using both of the methods, aiming for a comparison between them. Lupo et al. (2015) used these methods to encapsulate cocoa extract. They observed that external gelation produced smaller and harder particles than the internal method, due to the effect of syneresis and the rigidity of the formed wall, respectively, since the calcium migrates from the outside to the inside of the capsules. In another study, annato extract was encapsulated by external and internal gelations, with optimized conditions for both methods. Particles produced by external gelation presented higher encapsulation efficiency, larger size, and more homogeneous particle size than ones obtained by internal gelation (Naranjo-Durán et al. 2021). Rajmohan and Bellmer (2019) concluded that spirulina can be encapsulated by external or internal ionic gelation. Beads produced by external gelation exhibited a more uniform and homogeneous morphology, harder structure, and higher encapsulation efficiency than those obtained by internal gelation.

Microencapsulation by ionic gelation usually occurs in combination with other techniques, depending on several factors, such as particle characteristics, encapsulated compound, and wall material. Extrusion, atomization, and emulsification are the most common methods used. When ionic gelation is applied with extrusion, the mixture containing the polymer and the core material passes through a nozzle, forming droplets, which fall into a bath containing a solution that will cause the crosslinking (Tavlasoglu, Ozkan, and Capanoglu 2022; Cortés-Camargo et al. 2023). In addition to simple extrusion, more sophisticated equipment can be used for the particle formation, such as electrospinning (Kraithong et al. 2023), ultrasonic nozzle (Norcino et al. 2022), and vibration dripping extrusion (Sharma, Dash, and Badwaik 2022). Ionic gelation combined with extrusion is widely used for the microencapsulation of probiotics. Naissinger da Silva et al. (2021) encapsulated *Lactobacillus acidophilus* and *Bifidobacterium bifidum* using sodium alginate and applied the microparticles in butter. The authors observed that butter containing 5% and 10% of *Lactobacillus acidophilus* particles showed a count greater than 8 log CFU/g in storage for 45 days at 5°C, while free microorganisms in the butter did not present probiotic viability in any storage period.

When the objective is to obtain particles with smaller sizes than those produced by extrusion, atomization can be used. Some researchers have obtained particles containing probiotics by atomization followed by ionic gelation. Poletto et al. (2019) used rice bran, inulin, and Hi-maize in combination with sodium alginate to encapsulate *Lactobacillus acidophilus*. The authors produced a dispersion containing the wall materials and probiotics, that was atomized in the gelling bath, generating particles from 79.7 to 117.70 μm.

Regarding emulsification, this is a technique normally used as a pre-treatment in several cases. When the material to be encapsulated is hydrophobic, it must be homogenized with the polymer solution by emulsification, before the crosslinking (Feltre et al. 2020; Vakarelova et al. 2023; Odjo et al. 2022). Feltre et al. (2020) produced an emulsion with sodium alginate, gelatinized corn starch, and wheat germ oil and then atomized it in a $CaCl_2$ solution. An emulsion composed of an aqueous phase (a blend of gelatinized potato starch, gelatin, and alginate) and sunflower oil was obtained and then atomized or extruded in the gelling bath to produce emulsion-filled hydrogels (Silva et al., 2020).

Emulsification is also used when the core material is hydrophilic and it is desired to homogenize it to an oil to increase its encapsulation efficiency (Moura et al. 2022). Xu et al. (2019) dissolved curcumin in tea seed oil and then emulsified it with sodium alginate. After this process, the emulsion was dripped into a calcium chloride solution. The particles presented an encapsulation efficiency of 99.15% and a low release rate at pH 7.0. Probiotics also can be encapsulated by the combination of emulsification and ionic gelation. Farahmand et al. (2022) used emulsification as a pre-treatment to encapsulate *Lactobacillus plantarum* and *Bifidobacterium animalis* by millifluidic/direct gelation. In another study, *Lacticaseibacillus rhamnosus* GG was homogeneized with vegetable oil and a surfactant and then this mixture was dispersed in a sodium alginate solution, to the production of particles by ionic gelation that was applied in apple juice and yogurt (Romero-Chapol et al. 2022). Varela-Pérez et al. (2022) produced alginate particles loaded with *Lactobacillus gasseri* by ionic gelation using emulsification by two different homogenization methods: magnetic stirring and rotor-stator homogenizer as pre-treatment. The authors concluded that the cell viability obtained by the two methodologies was similar. The retention of cell viability after the gastrointestinal process was 100% to particles produced by magnetic stirring and 79.14% to that obtained with a rotor-stator homogenizer, while free cells did not present any viability.

Ionic gelation is a suitable method to encapsulate both hydrophobic and hydrophilic compounds, but it is more usually used in hydrophobic ones, due to the low solubility in the polysaccharides used as wall material. Some recent studies used ionic gelation for the microencapsulation of sesame oil (Cortés-Camargo et al. 2023), clove oil and thymol blend (Lim, Lee, and Lee 2023), ruta oil (Castillo et al. 2023), and microbial oil (Somacal et al. 2022), applying different polymers as wall materials.

Hydrophilic compounds also have been encapsulated by ionic gelation, but the methodology, in this case, presents some limitations since the polysaccharide solutions and the core materials are miscible with each other, which may hinder the retention of these compounds within the matrix. Some studies have proposed alternatives to avoid this problem, using methodologies such as emulsification, adsorption, and coating. To encapsulate anthocyanins from jussara extract, Da Silva Carvalho et al. (2019) used two strategies: adsorption and coating. First, the authors produced alginate beads without core, that adsorbed anthocyanins from concentrated jussara extract kept at pH 3. After this process, beads were coated with chitosan, gelatin, or whey protein concentrate solutions at acid pH, by electrostatic interaction. During 4 weeks of storage at 5° C, the gelatin-coated beads showed the lowest anthocyanin loss, while the uncoated beads presented the greatest loss. In another work, hibiscus extract was emulsified with rapessed oil and a surfactant and then, a double emulsion (w/o/w) was produced with pectin solution. Ionic gelation occurred with two different combined methods: extrusion and atomization. Atomized particles presented better encapsulation efficiency and antioxidant capacity, while particles produced by extrusion showed greater stability. The study concluded that ionic gelation, with both combined methods, was suitable to microencapsulate heat-sensitive compounds (de Moura et al. 2018). The emulsification process was also used previously the ionic gelation to encapsulate a hydrophilic compound by Yousefi et al. (2020). *Viola odorata L.* extract was first added to the alginate solution and then a W/O emulsion was produced with soybean oil. $CaCl_2$ solution was dripped in the emulsion, forming the microparticles, followed by chitosan coating. The encapsulation allowed the maintenance of the extract antioxidant activity during 4 weeks of storage, while the free extract had it decreased (Yousefi et al. 2020). The addition of the *stevia rebaudiana Bertoni* extract in the crosslinker bath was performed by Arriola et al. (2019), aiming to improve the entrapment efficiency of stevia phenolic compounds and avoid losses by diffusion. The beads obtained by the use of $CaCl_2$ containing dissolved stevia extract exhibited higher encapsulation efficiency, total phenolic content, and lower size when compared to the beads produced with only an aqueous solution of $CaCl_2$.

Although ionic gelation is a widely used technique and has several advantages, alginate matrices have a porous structure, which can cause exposure of the encapsulated compound and, consequently, its deterioration or release at an undesired time or place. To avoid or minimize this limitation, several studies have applied different materials as fillers in alginate matrices. Proteins (Silva et al. 2021; Mazza et al. 2023; Shakoor, Pamunuwa, and Karunaratne 2023) and polysaccharides, such as starch (Feltre et al. 2020; Poletto et al. 2019; Kamaldeen, Arianhy, and Yusufu 2020), mucilages

(Cortés-Camargo et al. 2023; Velázquez-Gutiérrez et al. 2021) and pectin (Tavlasoglu, Ozkan, and Capanoglu 2022; Norcino et al. 2022) are examples of materials used for this purpose. Huang et al. (2020) encapsulated probiotics (*Limosilactobacillus reuteri* TMW) using several different fillers combined with alginate (shellac, whey protein isolate, and sucrose). They observed that the addition of shellac in the composition of the particles improved microorganisms viability after freeze drying, simulated digestibility, heating (80 °C/ 26% RH), and ambient storage.

Another alternative to improve the properties of the particles and reduce the exposure of the encapsulated compound is to perform the coating of the particles produced by ionic gelation. A material widely used for coating is chitosan, which interacts electrostatically with sodium alginate (Ramos et al. 2018). The use of chitosan coating on alginate particles for encapsulation of *Saccharomyces boulardii* showed greater protection than uncoated particles for all studied storage temperatures (-18, 7, and 30 °C/ 120 days) (Santos and Machado 2020).

## Recent Food Applications of Encapsulated Probiotics by Ionic Gelation

The literature presents a wide variety of studies that show the benefits to health and well-being promoted by probiotic consumption. Several beneficial effects have been associated with the modulation of the intestinal microbiota (Echegaray et al. 2023; Yadav et al. 2013), competition and inhibition of pathogenic microorganisms; production of antibacterial and antioxidant compounds (Parlindungan, Dekiwadia, and Jones 2021; Zhou et al. 2022), beneficial action in reducing cholesterol (Jitpakdee et al. 2021; Majeed et al. 2019), combating diabetes (Y. Liu et al. 2022), obesity (Alard et al. 2021), modulation of immune responses (Lehtoranta, Latvala, and Lehtinen 2020; Maldonado Galdeano et al. 2019), among others.

However, to promote their benefits, probiotics need to resist adverse conditions during processing (temperature, pressure, oxygen, interaction with food matrices where they are inserted), transport, and storage, as well as drastic conditions during consumption (exposure to gastrointestinal fluids) (Liu et al., 2019). To exert their action, probiotics must reach the intestinal tract in sufficient amounts (from 6 to 8 log CFU/g or mL) (Chávarri et al. 2010). As already explored in this chapter, probiotic encapsulation by ionic gelation, and its combination with other encapsulation techniques, can improve the resistance of probiotic microorganisms, in addition to the potential to promote controlled delivery (Silva et al. 2018; Cezarino et al. 2022).

Figure 1 presents a scheme of the probiotic encapsulation process by the ionic gelation technique associated with atomization or extrusion. Considering that probiotics exert their action in the intestine, the particles must be designed to resist passage through gastric fluids and promote probiotics release in the intestinal tract. The technique uses surface-charged hydrocolloid solutions containing microbial cultures, which are extruded or atomized in an oppositely charged saline solution. Depending on the wall materials used, the gel is stable in acidic media but disintegrates in an alkaline environment, typical of the intestinal tract.

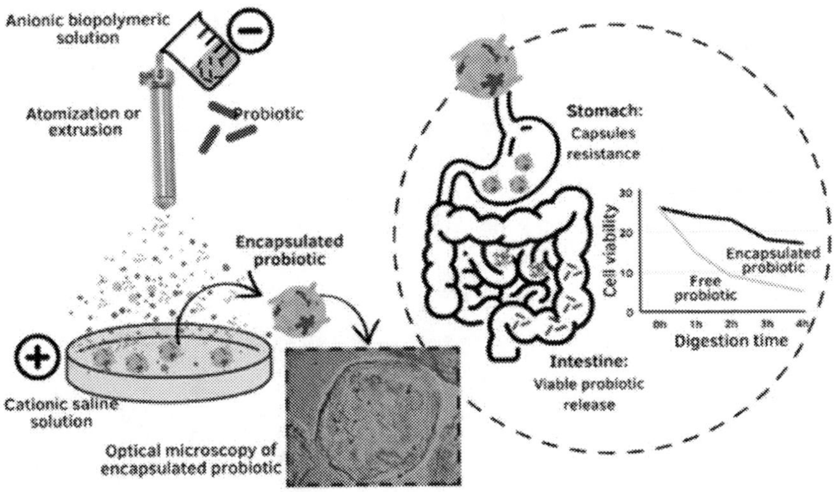


**Figure 1.** Encapsulation of probiotic cells by ionic gelation technique. During consumption, the particles must resist gastric conditions and disintegrate in the intestinal tract, releasing the probiotics.

Currently, healthy foods have shown a growing demand by consumers, which has driven the development of food products containing probiotics in recent years. A variety of foods with distinct characteristics in terms of composition, nutritional and sensory aspects, meeting specific diets and restrictions of consumers has emerged, enabling the widespread consumption of probiotics (Rodrigues et al. 2020; Feng et al. 2023). Dairy (Cezarino et al. 2022) and non-dairy products (Zhao et al. 2021), chocolates (Hossain et al. 2021), meat products (Bucheli, Fugaban, and Todorov 2022), food supplements (Putta et al. 2018) and many other products can be found. Consequently, the market for supplements and foods with probiotics addition has been growing. It is estimated that the global probiotics market reached US$ 58 billion in 2019 and the future estimates are US$ 85 billion by 2027 (Markets and Markets 2022). Figure 2 presents some possible applications of encapsulated probiotics in food summarized through examples presented by the literature (Granato et al. 2010; Khan et al. 2011; Rodrigues et al. 2020).

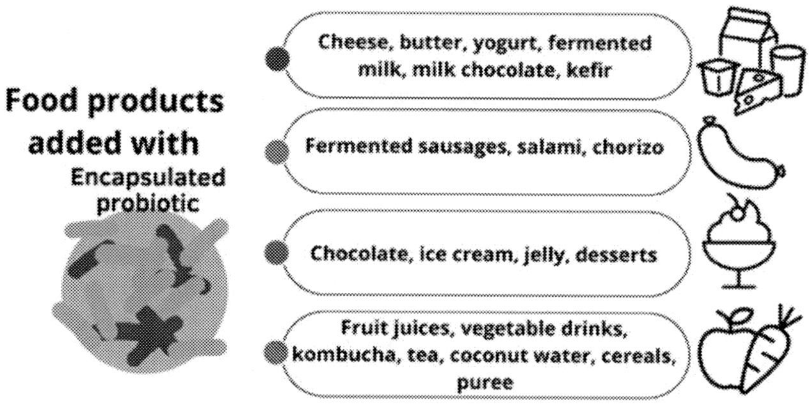


**Figure 2.** Food matrices are commonly used for probiotics addition.

## Dairy Products

Due to the physicochemical characteristics of dairy products, fermented or unfermented, including high availability of carbohydrates, proteins, lipids, vitamins, and interaction with other microorganisms, these foods are an excellent option to carry probiotic bacteria, being explored for a long time. The dairy products category includes products such as cheeses, butter, fermented beverages, and desserts (Rodrigues et al. 2020; Gullo and Zotta 2022).

During the processing of fermented dairy products, probiotics are often natural constituents, such as yogurt, a product traditionally obtained by fermentation of milk by the starter cultures *L. delbrueckii subsp. bulgaricus* and *S. thermophilus are* considered probiotics. However, the addition of other probiotic strains has been successfully used in the manufacture of yogurts, resulting in desirable functional and organoleptic characteristics (Gao et al. 2021).

Despite showing viability during the process and promoting desirable sensory transformations in yogurts, to ensure greater probiotic viability, many studies have investigated the potential of their microencapsulation. Aiming to increase *L. acidophilus* viability in yogurt and when exposed to the gastrointestinal tract (GIT), alginate particles were produced by ionic gelation with or without fructooligosaccharides (FOS) addition. Encapsulation provided greater probiotic viability to storage and protection when exposed to simulated GIT. The results reinforced the role of FOS as an alginate filler, acting as a substrate and consequently, contributing to better probiotic protection during yogurt storage (Silva et al. 2018).

Posteriorly, the same encapsulation systems were incorporated into liquid and frozen yogurt produced in laboratory and commercial Greek yogurt, for storage and *in vitro* simulated digestion evaluation. The highest viability during storage (28 days) was recorded for probiotics encapsulated with alginate and FOS added to liquid yogurt, followed by frozen and Greek yogurt (0.5; 2, and 2 logarithmic cycles reduction, respectively. The research highlighted the food matrix and storage temperature interference on the *L. acidophilus* viability. The microparticles improved cell viability under gastric conditions compared to free probiotics. However, the Greek yogurt probiotic cells were not viable during the GIT, reinforcing the influence of inhibitory substances present in commercial Greek yogurt on probiotic survival (Cezarino et al. 2022).

Cheeses represent food matrices that provide many favorable conditions for maintaining probiotic viability, such as higher pH (compared to other fermented dairy products), maturation processes, nutrient availability, and lower water activity, which promotes buffering, in addition to the low oxygen content. All these conditions allow better development, viability, and protection of probiotics. Adding encapsulated probiotics to cheeses may further improve their functionality (Rolim et al. 2020). For example, alginate particles produced by ionic gelation to protect *L. acidophilus* improved the survival of probiotics during storage and *in vitro* digestion, in addition to attributing desirable technological characteristics to goat ricotta, in which the particles were added (Lopes et al. 2021).

An extensive repertoire of dairy products added from encapsulation systems obtained by ionic gelation can be cited, including kefir (González-

Sánchez et al. 2010), butter (Naissinger da Silva, Tagliapietra, and dos Santos Richards 2021), ice cream (Zaeim et al. 2020), desserts (Karimi, Sekhavatizadeh, and Hosseinzadeh 2021), among others. Probiotics can be inserted into encapsulation systems alone or with prebiotics, combining encapsulation techniques or biopolymers, strategies that can increase their nutraceutical potential.

## Fruit and Vegetable Fermented Foods

To meet the different needs of consumers, research on probiotics incorporated in non-dairy products has grown significantly, targeting consumers with specific eating habits, such as vegetarians, lactose intolerant, or allergic to milk protein, who are looking for low-cholesterol and nutritious foods. Against this background, the food industry is focusing mainly on fruit and vegetable-based fermented foods (Kumar et al. 2022).

In the vegetable beverage sector, some challenges are faced for the insertion of probiotics, as sensory changes can occur, impacting characteristics such as aroma, flavor, appearance, and texture of beverages, due to the products generated by the probiotic metabolism (Marrero et al. 2019). In this sense, microencapsulation offers the possibility of increasing the functionality of these foods by probiotics addition, reducing the cell interaction with the food matrix, in addition to ensuring greater probiotic protection and viability (Pimentel et al. 2021).

The internal gelation method was used to encapsulate *B. bifidum* in alginate and k-carrageenan matrix. The particles were added to pasteurized grape juice. Compared to free cells (6.58 log CFU/mL), the particles improved the viability of probiotics (7.09 log CFU/mL) on storage and enhances the survival of cells during simulated gastric conditions. However, the authors pointed out that the viable probiotic count at 35 days was lower than the minimum recommended and some sensory changes were noted (Afzaal et al. 2020).

Alginate and vegetable oil capsules loaded with *L. gasseri* were obtained by emulsification and ionic gelation in addition to apple juice. Cell viability after three weeks of storage was 77% for capsules subjected to homogenization in magnetic stirring (AM) and 63.1% for free cells. At the end of the *in vitro* gastrointestinal simulation, free cells did not show viability, while the AM system showed 100% viability (Varela-Pérez et al. 2022).

Similar to fruit-based products, many studies reported the addition of probiotic particles in vegetable beverages, such as rice and soybean milk (Lopes et al. 2020). Alginate and alginate-coated chitosan particles were obtained for *L. acidophilus* protection by ionic gelation. Compared to free probiotic, microencapsulated *L. acidophilus* maintained adequate cell counts in rice and soybean milk during storage at 7 °C for 7 days (>7 logs CFU/mL) and simulated gastrointestinal digestion (>6 logs CFU/mL in the ileum) (Lopes et al. 2020). It is noteworthy that fruit and vegetable beverages have the potential to preserve probiotic viability, as they are a source of prebiotic compounds (Lillo-Pérez et al. 2021).

## Meat Products

Among meat products, sausages are potential candidates for probiotic insertion, as heat treatment is not used in most processing, which increases the probiotic survival chances. In addition to the functional characteristics, lactic acid bacteria have also been used to improve meat safety, as an effect on reducing the virulence of *Listeria* (Khan et al. 2011).

In fermented meat sausages, the quality of the final product is closely related to maturation, a processing stage that favors the development of microorganisms that provide the sensory, nutritional, and safety characteristics of the products. In these products, a high amount of microorganisms is present, among them, some bacterial strains that have been characterized as probiotics. However, incorporated probiotics must show adaptation to fermentation conditions and long-term storage, becoming dominant (Ruiz-Moyano et al. 2011). The encapsulation technique can overcome problems of low probiotic survival in fermented meat products.

Different strategies for the *L. plantarum* incorporation in chorizo sausages were investigated by (Cavalheiro et al. 2019). In addition to promoting fermentation in meat products, *L. plantarum* provides its probiotic benefits. Among the evaluated strategies (free cells, particles obtained by ionic gelation, and emulsion), the ionic gelation technique showed the best results for improving probiotic count during maturation, in addition to promoting desirable physical-chemical and sensory characteristics. Dry fermented sausages were also inoculated with *Enterococcus faecium* (free cell or encapsulated in alginate particles). The probiotic particles resulted in harder sausages, with higher protein and fat content, and lower moisture. However, no differences were observed in the viability of *E. faecium* added as free or encapsulated at 60 days of storage (Cavalheiro et al. 2021).

In meat products, *L. monocytogenes* causes listeriosis, a serious disease. The addition of bacteriocins in meat products has been investigated for conservation, but for its production, favorable conditions must be guaranteed for the producing bacteria. Barbosa, Todorov, Jurkiewicz, and Franco (2015) evaluated the addition of *Lb. curvatus* free and encapsulated into a calcium alginate matrix for the reduction of *L. monocytogenes* in salami during the manufacturing process. Encapsulated *Lb. curvatus* presented greater viability, however, promoted a similar reduction in *L. monocytogenes* counts in salami.

In summary, probiotic strains added to meat products must be strong competitors against the natural microflora of meat, provide desirable sensory transformations, combat pathogens, and, depending on the product, resist process conditions, such as temperature, in addition to being viable for exercise and its benefits during consumption. Encapsulation systems have been used for different purposes; however, the sector indicates that much research can be developed to obtain better and more functional products.

## Conclusion

Ionic gelation is a widely used microencapsulation technique. In this method, particles are formed through crosslinking between polysaccharide chains and oppositely charged ions.

Alginate is the most used polysaccharide for encapsulation by ionic gelation, however, it forms a porous network, which can reduce the protection of probiotics. Pectin and gellan are also employed in ionic gelation. To overcome the sensitivities of alginate particles, its combination with other polymers, fillers, or association with other encapsulation techniques has been investigated. Therefore, ionic gelation is a process that presents technical malleability.

Microencapsulation by atomization or extrusion followed by ionic gelation allows large-scale production, which makes it economically viable, offers stability to probiotics, and enables the incorporation of encapsulation systems in various products, especially in products that have a high-water amount, such as dairy products, fruit juices, vegetable beverages, meat sausages, desserts, among others.

The incorporation of probiotics into foods on an industrial scale involves microbiological, technological, and economic challenges. The sensitivity of probiotics to environmental conditions, such as thermal and osmotic stress, exposure to oxygen and acidic conditions, interaction with food matrices, as well as during their passage through the upper gastrointestinal tract, can be mentioned.

Many studies point to the potential of the ionic gelation technique to protect probiotics added to foods, emphasizing the viability of these encapsulated microorganisms under process, storage, and consumption conditions. These researches allowed that nowadays it is possible to consume these products and obtain their benefits.

## References

Afzaal, Muhammad, Farhan Saeed, Muhammad Saeed, Muhammad Azam, Shahzad Hussain, Abdellatif A Mohamed, Mohamed S Alamri, and Faqir Muhammad Anjum. 2020. "Survival and Stability of Free and Encapsulated Probiotic Bacteria under Simulated Gastrointestinal and Thermal Conditions." *International Journal of Food Properties* 23 (1): 1899–1912.https://doi.org/10.1080/10942912.2020.1826513

Alard, Jeanne, Benoit Cudennec, Denise Boutillier, Véronique Peucelle, Amandine Descat, Raphaël Decoin, Sarah Kuylle, Amin Jablaoui, Moez Rhimi, Isabelle Wolowczuk, Bruno Pot, Anne Tailleux, Emmanuelle Maguin, Sophie Holowacz, Corinne Grangette. 2021. "Multiple Selection Criteria for Probiotic Strains with High Potential for Obesity Management." *Nutrients* 13 (3). https://doi.org/10.3390/nu13030713

Arriola, Nathalia D Aceval, Peter Ian Chater, Matthew Wilcox, Luigi Lucini, Gabriele Rocchetti, Milene Dalmina, Jeffrey P Pearson, and Renata Dias de Mello Castanho Amboni. 2019. "Encapsulation of Stevia Rebaudiana Bertoni Aqueous Crude Extracts by Ionic Gelation – Effects of Alginate Blends and Gelling Solutions on the Polyphenolic Profile." *Food Chemistry* 275: 123–34. https://doi.org/10.1016/j.foodchem.2018.09.086

Barbosa, Matheus S, Svetoslav D Todorov, Cynthia H Jurkiewicz, and Bernadette D G M Franco. 2015. "Bacteriocin Production by Lactobacillus Curvatus MBSa2 Entrapped in Calcium Alginate during Ripening of Salami for Control of Listeria Monocytogenes." *Food Control* 47: 147–53. https://doi.org/10.1016/j.foodcont.2014.07.005

Bucheli, Jorge Enrique Vazquez, Joanna Ivy Irorita Fugaban, and Svetoslav Dimitrov Todorov. 2022. "Chapter 7 - Probiotics in Meat Products." In *Probiotics*, edited by Adriano Brandelli, 129–43. Academic Press. https://doi.org/10.1016/B978-0-323-85170-1.00009-9

Burey, P., B. R. Bhandari, T. Howes, and M. J. Gidley. 2008. "Hydrocolloid Gel Particles: Formation, Characterization, and Application." *Critical Reviews in Food Science and Nutrition* 48 (5): 361–77. https://doi.org/10.1080/10408390701347801

Castillo, Duvan Fernando, Rubén Albeiro Sánchez-Andica, Brayan Fernando Enriquez, Jaime Restrepo, and Martha Isabel Páez-Melo. 2023. "Encapsulation of Ruta Essential

Oil in Chitosan and Alginate Matrices as an Ecological Alternative for the Control of Nematodes." *Journal of Microencapsulation* 0 (0): 1–13. https://doi.org/10.1080/02652048.2023.2188939

Cavalheiro, Carlos Pasqualin, Claudia Ruiz-Capillas, Ana Maria Herrero, Francisco Jiménez-Colmenero, Tatiana Pintado, Cristiano Ragagnin de Menezes, and Leadir Lucy Martins Fries. 2019. "Effect of Different Strategies of Lactobacillus Plantarum Incorporation in Chorizo Sausages." *Journal of the Science of Food and Agriculture* 99 (15): 6706–12. https:/doi.org/10.1002/jsfa.9952

Cavalheiro, Carlos Pasqualin, Claudia Ruiz-Capillas, Ana Maria Herrero, and Tatiana Pintado. 2021. "Dry-Fermented Sausages Inoculated with Enterococcus Faecium CECT 410 as Free Cells or in Alginate Beads." *LWT* 139: 110561. https://doi.org/10.1016/j.lwt.2020.110561

Cezarino, Eliza Cristina, Karen Cristina Guedes Silva, Flávia Souza Almeida, and Ana Carla Kawazoe Sato. 2022. "Stability and Viability of Synbiotic Microgels Incorporated into Liquid, Greek and Frozen Yogurts." *Journal of Food Science* 87 (4): 1796–1809.https:/doi.org/10.1111/1750-3841.16107

Chai, Jingjing, Ping Jiang, Pengjie Wang, Yumeng Jiang, Dan Li, Weier Bao, Bingxue Liu, Bin, Liu, Liyun, Zhao, Willem, Norde, Qipeng, Yuan, Fazheng, Ren, Yuan Li. 2018. "The Intelligent Delivery Systems for Bioactive Compounds in Foods: Physicochemical and Physiological Conditions, Absorption Mechanisms, Obstacles, and Responsive Strategies." *Trends in Food Science & Technology* 78: 144–54. https://doi.org/10.1016/j.tifs.2018.06.003

Chávarri, María, Izaskun Marañón, Raquel Ares, Francisco C Ibáñez, Florencio Marzo, and María del Carmen Villarán. 2010. "Microencapsulation of a Probiotic and Prebiotic in Alginate-Chitosan Capsules Improves Survival in Simulated Gastro-Intestinal Conditions." *International Journal of Food Microbiology* 142 (1): 185–89. https://doi.org/10.1016/j.ijfoodmicro.2010.06.022

Chun, Honam, Cheol-Hyun Kim, and Young-Hee Cho. 2014. "Microencapsulation of Lactobacillus Plantarum DKL 109 Using External Ionic Gelation Method." *Korean Journal for Food Science of Animal Resources* 34 (5): 692–99. https://doi.org/10.5851/kosfa.2014.34.5.692.

Cortés-Camargo, Stefani, Angélica Román-Guerrero, Jose Alvarez-Ramirez, Erik Alpizar-Reyes, Sandra Karina Velázquez-Gutiérrez, and César Pérez-Alonso. 2023. "Microstructural Influence on Physical Properties and Release Profiles of Sesame Oil Encapsulated into Sodium Alginate-Tamarind Mucilage Hydrogel Beads." *Carbohydrate Polymer Technologies and Applications* 5. https://doi.org/10.1016/j.carpta.2023.100302.

Silva Carvalho, Ana Gabriela da, Mariana Teixeira da Costa Machado, Helena Dias de Freitas Queiroz Barros, Cinthia Baú Betim Cazarin, Mário Roberto Maróstica Junior, and Miriam Dupas Hubinger. 2019. "Anthocyanins from Jussara (Euterpe edulis Martius) Extract Carried by Calcium Alginate Beads Pre-Prepared Using Ionic Gelation." *Powder Technology* 345: 283–91. https://doi.org/10.1016/j.powtec.2019.01.016.

Dickinson, Eric. 2003. "Hydrocolloids at Interfaces and the Influence on the Properties of Dispersed Systems." *Food Hydrocolloids* 17 (1): 25–39. https://doi.org/10.1016/S0268-005X(01)00120-5

Echegaray, Noemí, Birsen Yilmaz, Heena Sharma, Manoj Kumar, Mirian Pateiro, Fatih Ozogul, and Jose Manuel Lorenzo. 2023. "A Novel Approach to Lactiplantibacillus Plantarum: From Probiotic Properties to the Omics Insights." *Microbiological Research* 268: 127289. https://doi.org/10.1016/j.micres.2022.127289.

Farahmand, Atefeh, Behrouz Ghorani, Bahareh Emadzadeh, Mahboobe Sarabi-Jamab, Maryam Emadzadeh, Atena Modiri, and Nick Tucker. 2022. "Millifluidic-Assisted Ionic Gelation Technique for Encapsulation of Probiotics in Double-Layered Polysaccharide Structure." *Food Research International* 160: 111699. https://doi.org/10.1016/j.foodres.2022.111699.

Feltre, Gabriela, Tanara Sartori, Klycia F C Silva, Gustavo C Dacanal, Florencia C Menegalli, and Miriam D Hubinger. 2020. "Encapsulation of Wheat Germ Oil in Alginate-Gelatinized Corn Starch Beads: Physicochemical Properties and Tocopherols' Stability." *Journal of Food Science* 85 (7): 2124–33. https://doi.org/10.1111/1750-3841.15316.

Feng, Shucen, Haitao Wang, Xinping Lin, Huipeng Liang, Sufang Zhang, Yingxi Chen, and Chaofan Ji. 2023. "Probiotic Properties of Lactobacillus Plantarum and Application in Prebiotic Gummies." *LWT* 174: 114357. https://doi.org/10.1016/j.lwt.2022.114357.

Funami, Takahiro, Yapeng Fang, Sakie Noda, Sayaka Ishihara, Makoto Nakauma, Kurt I Draget, Katsuyoshi Nishinari, and Glyn O Phillips. 2009. "Rheological Properties of Sodium Alginate in an Aqueous System during Gelation about Supermolecular Structures and Ca 2 þ Binding." *Food Hydrocolloids* 23: 1746–55.https://doi.org/10.1016/j.foodhyd.2009.02.014

Gao, Jie, Xiyu Li, Guohua Zhang, Faizan Ahmed Sadiq, Jesus Simal-Gandara, Jianbo Xiao, and Yaxin Sang. 2021. "Probiotics in the Dairy Industry—Advances and Opportunities." *Comprehensive Reviews in Food Science and Food Safety* 20 (4): 3937–82. https://doi.org/10.1111/1541-4337.12755.

González-Sánchez, F, A Azaola, G F Gutiérrez-López, and H Hernández-Sánchez. 2010. "Viability of Microencapsulated Bifidobacterium Animalis Ssp. Lactis BB12 in Kefir during Refrigerated Storage." *International Journal of Dairy Technology* 63 (3): 431–36. https://doi.org/10.1111/j.1471-0307.2010.00604.x.

Granato, Daniel, Gabriel F Branco, Filomena Nazzaro, Adriano G Cruz, and José A F Faria. 2010. "Functional Foods and Nondairy Probiotic Food Development: Trends, Concepts, and Products." *Comprehensive Reviews in Food Science and Food Safety* 9 (3): 292–302. https://doi.org/10.1111/j.1541-4337.2010.00110.x.

Grant, Gregor T., Edwin R. Morris, David A. Rees, Peter J. C. Smith, and David Thom. 1973. "Biological Interactions between Polysaccharides and Divalent Cations: The Egg-Box Model." *FEBS Letters* 32 (1): 195–98. https://doi.org/10.1002/rmv.519

Gullo, Maria, and Teresa Zotta. 2022. "Chapter 18 - Probiotics in Dairy Products: Microencapsulation and Delivery." In *Advances in Dairy Microbial Products*, edited by Joginder Singh and Ashish Vyas, 271–85. Woodhead Publishing. / https://doi.org/10.1016/B978-0-323-85793-2.00032-1.

Helgerud, Trond, G Olav, Therese Fjæreide, Peder O Andersen, and Christian K Larsen. 2010. "Alginates." In *Food Stabilisers, Thickeners and Gelling Agents*, edited by A Imeson, 50–72. New Jersey, USA: Blackwell Publishing Ltd.

Hossain, Md Nur, Chaminda Senaka Ranadheera, Zhongxiang Fang, and Said Ajlouni. 2021. "Impact of Encapsulating Probiotics with Cocoa Powder on the Viability of Probiotics during Chocolate Processing, Storage, and in Vitro Gastrointestinal Digestion." *Journal of Food Science* 86 (5): 1629–41. https://doi.org/10.1111/1750-3841.15695.

Huang, Xue, Michael Gänzle, Hui Zhang, Meng Zhao, Yapeng Fang, and Katsuyoshi Nishinari. 2020. "Microencapsulation of Probiotic Lactobacilli with Shellac as Moisture Barrier and to Allow Controlled Release." *Journal of the Science of Food and Agriculture* 101 (2): 726–34. https://doi.org/10.1002/jsfa.10685.

Jitpakdee, Jirayu, Duangporn Kantachote, Hiroshi Kanzaki, and Teruhiko Nitoda. 2021. "Selected Probiotic Lactic Acid Bacteria Isolated from Fermented Foods for Functional Milk Production: Lower Cholesterol with More Beneficial Compounds." *LWT* 135: 110061https:/doi.org/10.1016/j.lwt.2020.110061

Kamaldeen, O S, C C Arianhy, and M I Yusufu. 2020. "Application of Soy Protein Isolate and Cassava Starch-Based Film Solutions as Matrix for Ionic Encapsulation of Carrot Powders." *Journal of Food Science and Technology* 57 (11): 4171–81. https://doi.org/10.1007/s13197-020-04455-w

Karimi, Masoumeh, Seyed Saeed Sekhavatizadeh, and Saeid Hosseinzadeh. 2021. "Milk Dessert Containing Lactobacillus Reuteri (ATCC 23272) Encapsulated with Sodium Alginate, Ferula Assa-Foetida and Zedo (Amygdalus Scoparia) Gum as Three Layers of Wall Materials." *Food and Bioproducts Processing* 127: 244–54. https://doi.org/10.1016/j.fbp.2021.03.003.

Khan, Muhammad Issa, Muhammad Sajid Arshad, Faqir Muhammad Anjum, Ayesha Sameen, Aneeq-ur-Rehman, and Waqas Tariq Gill. 2011. "Meat as a Functional Food with Special Reference to Probiotic Sausages." *Food Research International* 44 (10): 3125–33. https://doi.org/10.1016/j.foodres.2011.07.033.

Kraithong, Supaluck, Shahid Ahmed Junejo, Yi Jiang, Bin Zhang, and Qiang Huang. 2023. "Effects of Pectin-Calcium Matrices on Controlling in Vitro Digestion of Normal Maize Starch." *Food Hydrocolloids* 140: 108575. https://doi.org/10.1016/j.foodhyd.2023.108575.

Kumar, Sachin, Gurdeep Rattu, Swati Mitharwal, Abhishek Chandra, Sourabh Kumar, Aman Kaushik, Vijendra Mishra, and Prabhat K Nema. 2022. "Trends in Non-Dairy-Based Probiotic Food Products: Advances and Challenges." *Journal of Food Processing and Preservation* 46 (9): e16578. https://doi.org/10.1111/jfpp.16578.

Lehtoranta, Liisa, Sinikka Latvala, and Markus J Lehtinen. 2020. "Role of Probiotics in Stimulating the Immune System in Viral Respiratory Tract Infections: A Narrative Review." *Nutrients* 12 (10). https://doi.org/10.3390/nu12103163

Lillo-Pérez, Siegried, María Guerra-Valle, Patricio Orellana-Palma, and Guillermo Petzold. 2021. "Probiotics in Fruit and Vegetable Matrices: Opportunities for Nondairy Consumers." *LWT* 151: 112106. https://doi.org/10.1016/j.lwt.2021.112106.

Lim, Do Yeon, Ji-Soo Lee, and Hyeon Gyu Lee. 2023. "Nano-Encapsulation of a Combination of Clove Oil and Thymol and Their Application in Fresh-Cut Apples and

Raw Minced Beef." *Food Control* 148. https://doi.org/10.1016/j.foodcont.2023.109683.

Liu, Huan, Steve W Cui, Maoshen Chen, Yue Li, Rong Liang, Feifei Xu, and Fang Zhong. 2019. "Protective Approaches and Mechanisms of Microencapsulation to the Survival of Probiotic Bacteria during Processing, Storage and Gastrointestinal Digestion: A Review." *Critical Reviews in Food Science and Nutrition* 59 (17): 2863–78. https://doi.org/10.1080/10408398.2017.1377684

Liu, Y, S Zheng, J Cui, T Guo, and J Zhang. 2022. "Lactiplantibacillus Plantarum Y15 Alleviate Type 2 Diabetes in Mice via Modulating Gut Microbiota and Regulating NF-KB and Insulin Signaling Pathway." *Brazilian Journal of Microbiology* 53 (2): 935–45. https://doi.org/10.1007/s42770-022-00686-5

Lopes, Laênia Angélica Andrade, Rafaela de Siqueira Ferraz Carvalho, Nereide Stela Santos Magalhães, Marta Suely Madruga, Ana Julia Alves Aguiar Athayde, Isabella Araújo Portela, Carlos Eduardo Barão, Tatiana Colombo Pimentel, Marciane Magnani, and Thayza Christina Montenegro Stamford. 2020. "Microencapsulation of Lactobacillus Acidophilus La-05 and Incorporation in Vegan Milks: Physicochemical Characteristics and Survival during Storage, Exposure to Stress Conditions, and Simulated Gastrointestinal Digestion." *Food Research International* 135: 109295.https:/doi.org/10.1016/j.foodres.2020.109295

Lopes, Laênia Angélica Andrade, Tatiana Colombo Pimentel, Rafaela de Siqueira Ferraz Carvalho, Marta Suely Madruga, Mércia de Sousa Galvão, Taliana Kenia Alencar Bezerra, Carlos Eduardo Barão, Marciane Magnani, and Thayza Christina Montenegro Stamford. 2021. "Spreadable Goat Ricotta Cheese Added with Lactobacillus Acidophilus La-05: Can Microencapsulation Improve the Probiotic Survival and the Quality Parameters?" *Food Chemistry* 346: 128769. https://doi.org/10.1016/j.foodchem.2020.128769.

Lopes Neto, José Honório Pereira, Marília Cícera Gomes dos Santos, Kerolayne Santos Leite, Liliane Andrade da Silva, Maria Isabel Ferreira Campos, Emmanuel Souza da Silveira, João Bosco Sousa Amaral, Marta Suely Madruga, Ana Luiza Mattos Braga, and Haíssa Roberta Cardarelli. 2021. "Development and Characterization of Lactobacillus Acidophilus (LA-3) Microparticles with Reducing Substances and Its Addition to Reino Cheese." *LWT* 143: 111083. https://doi.org/10.1016/j.lwt.2021.111083.

Lu, Wangxing, Xing Yang, Jian Shen, Zongze Li, Songwen Tan, Wenjie Liu, and Zeneng Cheng. 2021. "Choosing the Appropriate Wall Materials for Spray-Drying Microencapsulation of Natural Bioactive Ingredients: Taking Phenolic Compounds as Examples." *Powder Technology* 394: 562–74. https:/doi.org/10.1016/j.powtec.2021.08.082.

Lupo, Bryshila, Alicia Maestro, José M. Gutiérrez, and Carmen González. 2015. "Characterization of Alginate Beads with Encapsulated Cocoa Extract To Prepare Functional Food: Comparison of Two Gelation Mechanisms." *Food Hydrocolloids* 49: 25–34. https://doi.org/10.1016/j.foodhyd.2015.02.023

Majeed, Muhammed, Shaheen Majeed, Kalyanam Nagabhushanam, Sivakumar Arumugam, Kirankumar Beede, and Furqan Ali. 2019. "Evaluation of the in Vitro Cholesterol-Lowering Activity of the Probiotic Strain Bacillus CoagulansMTCC

5856." *International Journal of Food Science \& Technology* 54 (1): 212–20. https:/doi.org/10.1111/ijfs.13926

Maldonado Galdeano, C, S I Cazorla, J M Lemme Dumit, E Vélez, and G Perdigón. 2019. "Beneficial Effects of Probiotic Consumption on the Immune System." *Annals of Nutrition and Metabolism* 74 (2): 115–24. https://doi.org/10.1159/000496426

Markets, and Markets. 2022. "Probiotic Market by Ingredient (Bacteria and Yeast), Product Type (Functional Food & Beverages, Dietary Supplements, and Feed), End User (Human and Animal), Distribution Channel and Region - Global Industry Analysis, Size, Share, Growth, Regional Outlook." *Markets and Markets.* 2022. https://www.marketsandmarkets.com/Market-Reports/probiotic-market-advanced-technologies-and-global-market-69.html#:~:text=With a CAGR of 8.1, it is potential for future investment.

Marrero, Silvia Carballo, Alejandro Martínez-Rodríguez, Salvador E Maestre Pérez, and Soledad Prats Moya. 2019. "2 - New Trends and Applications in Fermented Beverages." In *Fermented Beverages*, edited by Alexandru Mihai Grumezescu and Alina Maria Holban, 31–66. Woodhead Publishing. https://doi.org/10.1016/B978-0-12-815271-3.00002-6.

Mazza, Karen Elbert Leal, André Mesquita Magalhães Costa, Janine Passos Lima da Silva, Daniela Sales Alviano, Humberto Ribeiro Bizzo, and Renata Valeriano Tonon. 2023. "Microencapsulation of Marjoram Essential Oil as a Food Additive Using Sodium Alginate and Whey Protein Isolate." *International Journal of Biological Macromolecules* 233.https://doi.org/10.1016/j.ijbiomac.2023.123478

Moura, Sílvia Cristina Sobottka Rolim de, Carolina L Berling, Sílvia P M Germer, Izabela D Alvim, and Míriam D Hubinger. 2018. "Encapsulating Anthocyanins from Hibiscus Sabdariffa L. Calyces by Ionic Gelation: Pigment Stability during Storage of Microparticles." *Food Chemistry* 241: 317–27. https://doi.org/10.1016/j.foodchem.2017.08.095

Moura, Sílvia Cristina Sobottka Rolim de, Gabriela Nicoluci Schettini, Darlila Aparecida Gallina, Izabela Dutra Alvim, and Miriam Dupas Hubinger. 2022. "Microencapsulation of Hibiscus Bioactives and Its Application in Yogurt." *Journal of Food Processing and Preservation* 46 (4): e16468. https://doi.org/10.1111/jfpp.16468.

Naissinger da Silva, Maritiele, Bruna Lago Tagliapietra, and Neila Silvia Pereira dos Santos Richards. 2021. "Encapsulation, Storage Viability, and Consumer Acceptance of Probiotic Butter." *LWT* 139: 110536. https://doi.org/10.1016/j.lwt.2020.110536.

Naranjo-Durán, Ana María, Julián Quintero-Quiroz, John Rojas-Camargo, and Gelmy Luz Ciro-Gómez. 2021. "Modified-Release of Encapsulated Bioactive Compounds from Annatto Seeds Produced by Optimized Ionic Gelation Techniques." *Scientific Reports* 11 (1): 1317. https://doi.org/10.1038/s41598-020-80119-1

Norcino, Laís Bruno, Juliana Farinassi Mendes, Jayne de Abreu Figueiredo, Natália Leite Oliveira, Diego Alvarenga Botrel, and Luiz Henrique Capparelli Mattoso. 2022. "Development of Alginate/Pectin Microcapsules by a Dual Process Combining Emulsification and Ultrasonic Gelation for Encapsulation and Controlled Release of Anthocyanins from Grapes (Vitis Labrusca L.)." *Food Chemistry* 391: 133256. https://doi.org/10.1016/j.foodchem.2022.133256

Odjo, Kabirou, Qais Ali Al-Maqtari, Hang Yu, Yunfei Xie, Yahui Guo, Mi Li, Yuhang Du, Kunfeng Liu, Yulun Chen, and Weirong Yao. 2022. "Preparation and Characterization of Chitosan-Based Antimicrobial Films Containing Encapsulated Lemon Essential Oil by Ionic Gelation and Cranberry Juice." *Food Chemistry* 397: 133781. https://doi.org/10.1016/j.foodchem.2022.133781.

Onsoyen, E. 1997. "Alginates." In *Thickening and Gelling Agents for Food*, edited by A. Imeson, 2nd ed., 22–44. London: Chapman & Hall.

Parlindungan, Elvina, Chaitali Dekiwadia, and Oliver A H Jones. 2021. "Factors That Influence Growth and Bacteriocin Production in Lactiplantibacillus Plantarum B21." *Process Biochemistry* 107: 18–26. https://doi.org/10.1016/j.procbio.2021.05.009.

Pimentel, Tatiana Colombo, Whyara Karoline Almeida da Costa, Carlos Eduardo Barão, Michele Rosset, and Marciane Magnani. 2021. "Vegan Probiotic Products: A Modern Tendency or the Newest Challenge in Functional Foods." *Food Research International* 140: 110033.https://doi.org/10.1016/j.foodres.2020.110033.

Poletto, Gabriela, Greice Carine Raddatz, Alexandre José Cichoski, Leila Queiroz Zepka, Eduardo Jacob Lopes, Juliano Smanioto Barin, Roger Wagner, and Cristiano Ragagnin de Menezes. 2019. "Study of Viability and Storage Stability of Lactobacillus Acidophilus When Encapsulated with the Prebiotics Rice Bran, Inulin and Hi-Maize." *Food Hydrocolloids* 95: 238–44. https://doi.org/10.1016/j.foodhyd.2019.04.049.

Poncelet, D., B. Poncelet De Smet, C. Beaulieu, M. L. Huguet, A. Fournier, and R. J. Neufeld. 1992. "Production of Alginate Beads by Emulsification/Internal Gelation. I. Methodology." *Applied Microbiology and Biotechnology* 38 (4): 39–45. https://doi.org/10.1007/BF00164768.

Puguan, John Marc C, Xiaohong Yu, and Hern Kim. 2014. "Characterization of Structure, Physico-Chemical Properties and Diffusion Behavior of Ca-Alginate Gel Beads Prepared by Different Gelation Methods." *Journal of Colloid and Interface Science* 432: 109–16. https://doi.org/10.1016/j.jcis.2014.06.048.

Putta, Swathi, Nagendra S Yarla, Dhananjaya B Lakkappa, Sarat B Imandi, Rama Rao Malla, Amajala K Chaitanya, Brahma P V Chari, Silas Saka, Rama R Vechalapu, Mohammad Amjad Kamal, Vadim V. Tarasov, Vladimir N. Chubarev, Korada Siva Kumar, Gjumrakch Aliev. 2018. "Chapter 2 - Probiotics: Supplements, Food, Pharmaceutical Industry." In *Therapeutic, Probiotic, and Unconventional Foods*, edited by Alexandru Mihai Grumezescu and Alina Maria Holban, 15–25. Academic Press. https://doi.org/10.1016/B978-0-12-814625-5.00002-9.

Rajmohan, Deepak, and Danielle Bellmer. 2019. "Characterization of Spirulina-Alginate Beads Formed Using Ionic Gelation." Edited by Salam A Ibrahim. *International Journal of Food Science* 2019: 7101279. https://doi.org/10.1155/2019/7101279

Ramos, Philippe E, Miguel A Cerqueira, José A Teixeira, and António A Vicente. 2018. "Physiological Protection of Probiotic Microcapsules by Coatings." *Critical Reviews in Food Science and Nutrition* 58 (11): 1864–77. https://doi.org/10.1080/10408398.2017.1289148.

Rodrigues, F J, M F Cedran, J L Bicas, and H H Sato. 2020. "Encapsulated Probiotic Cells: Relevant Techniques, Natural Sources as Encapsulating Materials and Food

Applications – A Narrative Review." *Food Research International* 137: 109682. https://doi.org/10.1016/j.foodres.2020.109682.

Rolim, Fernanda R L, Oliveiro C Freitas Neto, Maria Elieidy G Oliveira, Celso J B Oliveira, and Rita C R E Queiroga. 2020. "Cheeses as Food Matrixes for Probiotics: In Vitro and in Vivo Tests." *Trends in Food Science & Technology* 100: 138–54. https://doi.org/10.1016/j.tifs.2020.04.008.

Romero-Chapol, Oscar O, Abigail Varela-Pérez, Ana G Castillo-Olmos, Hugo S García, Jaspreet Singh, Pedro J García-Ramírez, Rubí Viveros-Contreras, Claudia Y Figueroa-Hernández, and Cynthia Cano-Sarmiento. 2022. "Encapsulation of Lacticaseibacillus Rhamnosus GG: Probiotic Survival, In Vitro Digestion and Viability in Apple Juice and Yogurt." *Applied Sciences* 12 (4). https://doi.org/10.3390/app12042141.

Ruiz-Moyano, Santiago, Alberto Martín, María José Benito, Alejandro Hernández, Rocio Casquete, and María de Guia Córdoba. 2011. "Application of Lactobacillus Fermentum HL57 and Pediococcus Acidilactici SP979 as Potential Probiotics in the Manufacture of Traditional Iberian Dry-Fermented Sausages." *Food Microbiology* 28 (5): 839–47. https://doi.org/10.1016/j.fm.2011.01.006.

Santos, Matheus A S, and Mariana T C Machado. 2020. "Coated Alginate–Chitosan Particles to Improve the Stability of Probiotic Yeast." *International Journal of Food Science \& Technology*, 1–10. https://doi.org/10.1111/ijfs.14829.

Shakoor, I Farrah, Geethi K Pamunuwa, and D Nedra Karunaratne. 2023. "Efficacy of Alginate and Chickpea Protein Polymeric Matrices in Encapsulating Curcumin for Improved Stability, Sustained Release and Bioaccessibility." *Food Hydrocolloids for Health* 3: 100119. https://doi.org/10.1016/j.fhfh.2023.100119.

Sharma, Maanas, Kshirod K Dash, and Laxmikant S Badwaik. 2022. "Physicochemical and Release Behaviour of Phytochemical Compounds Based on Black Jamun Pulp Extracts-Filled Alginate Hydrogel Beads through Vibration Dripping Extrusion." *International Journal of Biological Macromolecules* 194: 715–25. https://doi.org/10.1016/j.ijbiomac.2021.11.116.

Silva, Karen Cristina Guedes, Ana Isabel Bourbon, Lorenzo Pastrana, and Ana Carla Kawazoe Sato. 2020. "Emulsion-Filled Hydrogels for Food Applications: Influence of PH on Emulsion Stability and a Coating on Microgel Protection." *Food Funct.* 11 (9): 8331–41. https://doi.org/10.1039/D0FO01198C.

Silva, Karen Cristina Guedes, Eliza Cristina Cezarino, Mariano Michelon, and Ana Carla Kawazoe Sato. 2018. "Symbiotic Microencapsulation to Enhance Lactobacillus Acidophilus Survival." *LWT - Food Science and Technology* 89 (November 2017): 503–9. https://doi.org/10.1016/j.lwt.2017.11.026

Silva, Karen Cristina Guedes, Gabriela Feltre, Miriam Dupas Hubinger, and Ana Carla Kawazoe Sato. 2021. "Protection and Targeted Delivery of β-Carotene by Starch-Alginate-Gelatin Emulsion-Filled Hydrogels." *Journal of Food Engineering* 290: 110205.https:/doi.org/10.1016/j.jfoodeng.2020.110205

Smrdel, Polona, Marija Bogataj, Anamarija Zega, Odon Planinšek, and Aleš Mrhar. 2008. "Shape Optimization and Characterization of Polysaccharide Beads Prepared by Ionotropic Gelation." *Journal of Microencapsulation* 25 (2): 90–105. https://doi.org/10.1080/02652040701776109.

Somacal, Simara, Sabrina Somacal, Vandré Sonza Pinto, Cassandra de Deus, Raquel Guidetti Vendruscolo, Tielle Moraes de Almeida, Roger Wager, Márcio Antônio Mazutti, and Cristiano Ragagnin de Menezes. 2022. "Strategy to Increase the Lipid Stability of the Microbial Oil Produced by Umbelopsis Isabellina for Food Purposes: Use of Microencapsulation by External Ionic Gelation." *Food Research International* 152: 110907. https://doi.org/10.1016/j.foodres.2021.110907.

Tavlasoglu, Melda, Gulay Ozkan, and Esra Capanoglu. 2022. "Entrapment of Black Carrot Anthocyanins by Ionic Gelation: Preparation, Characterization, and Application as a Natural Colorant in Yoghurt." *ACS Omega* 7 (36): 32481 – 32488. https://doi.org/10.1021/acsomega.2c03962.

Vakarelova, Martina, Francesca Zanoni, Giulia Donà, Ilaria Fierri, Roberto Chignola, Sara Gorrieri, and Gianni Zoccatelli. 2023. "Microencapsulation of Astaxanthin by Ionic Gelation: Effect of Different Gelling Polymers on the Carotenoid Load, Stability and Bioaccessibility." *International Journal of Food Science and Technology* 58 (5): 2489 – 2497. https://doi.org/10.1111/ijfs.16389.

Varela-Pérez, Abigail, Oscar O Romero-Chapol, Ana G Castillo-Olmos, Hugo S García, Mirna L Suárez-Quiroz, Jaspreet Singh, Claudia Y Figueroa-Hernández, Rubí Viveros-Contreras, and Cynthia Cano-Sarmiento. 2022. "Encapsulation of Lactobacillus Gasseri: Characterization, Probiotic Survival, In Vitro Evaluation and Viability in Apple Juice." *Foods* 11 (5). https://doi.org/10.3390/foods11050740

Velázquez-Gutiérrez, S K, E Alpizar-Reyes, A Y Guadarrama-Lezama, J G Báez-González, J Alvarez-Ramírez, and C Pérez-Alonso. 2021. "Influence of the Wall Material on the Moisture Sorption Properties and Conditions of Stability of Sesame Oil Hydrogel Beads by Ionic Gelation." *LWT* 140: 110695. https://doi.org/10.1016/j.lwt.2020.110695.

Xu, Wei, Lu Huang, Weiping Jin, Peipei Ge, Bakht Ramin Shah, Dandan Zhu, and Junxiang Jing. 2019. "Encapsulation and Release Behavior of Curcumin Based on Nanoemulsions-Filled Alginate Hydrogel Beads." *International Journal of Biological Macromolecules* 134: 210–15. https://doi.org/10.1016/j.ijbiomac.2019.04.200.

Yadav, Ashok Kumar, Ashish Tyagi, Jai Kumar Kaushik, Asha Chandola Saklani, Sunita Grover, and Virender Kumar Batish. 2013. "Role of Surface Layer Collagen Binding Protein from Indigenous Lactobacillus Plantarum 91 in Adhesion and Its Anti-Adhesion Potential against Gut Pathogen." *Microbiological Research* 168 (10): 639–45. https://doi.org/10.1016/j.micres.2013.05.003

Yousefi, Mojtaba, Elham Khanniri, Mahdi Shadnoush, Nasim Khorshidian, and Amir M Mortazavian. 2020. "Development, Characterization and in Vitro Antioxidant Activity of Chitosan-Coated Alginate Microcapsules Entrapping Viola Odorata Linn. Extract." *International Journal of Biological Macromolecules* 163: 44–54. https://doi.org/10.1016/j.ijbiomac.2020.06.250.

Zaeim, Davood, Mahboobe Sarabi-Jamab, Behrouz Ghorani, Rassoul Kadkhodaee, Weilin Liu, and R Hans Tromp. 2020. "Microencapsulation of Probiotics in Multi-Polysaccharide Microcapsules by Electro-Hydrodynamic Atomization and Incorporation into Ice-Cream Formulation." *Food Structure* 25: 100147. https://doi.org/10.1016/j.foostr.2020.100147.

Zhao, Chu-Min, Ting Du, Ping Li, Xin-Jun Du, and Shuo Wang. 2021. "Production and Characterization of a Novel Low-Sugar Beverage from Red Jujube Fruits and Bamboo Shoots Fermented with Selected Lactiplantibacillus Plantarum." *Foods* 10 (7). https://doi.org/10.3390/foods10071439

Zhou, Y, W Gong, C Xu, Z Zhu, Y Peng, and C Xie. 2022. "Probiotic Assessment and Antioxidant Characterization of Lactobacillus Plantarum GXL94 Isolated from Fermented Chili." *Frontiers in Microbiology* 13: 997940. https://doi.org/10.3389/fmicb.2022.997940.

# Chapter 5

# Microencapsulation by Spray Drying

**Nataly de Almeida Costa***
Department of Food Technology (DTA), Federal University of Viçosa (UFV), Viçosa, Minas Gerais, Brazil

## Abstract

The microencapsulation of probiotics is an efficient alternative to promote cell protection during adverse conditions that may affect the viability and delivery of these microorganisms. Among the techniques used, microencapsulation by spray drying stands out for being an efficient technique, reproducible on an industrial scale, and capable of producing microparticles in a continuous flow. However, a high temperature used in the process can influence the viability of the probiotics present in the microparticles. Thus, the influence of the high temperature of the drying process will depend on the type of wall material used or their combination. In this chapter, the microencapsulation of probiotics by spray drying was described through the history of the technique, the description of the steps of the process, and the description of important points of the equipment such as the drying chamber, atomizer nozzle, and cyclone. In addition, some characteristics of microparticles were described, advantages and preservation of the technique, and different studies that use microencapsulation by spray drying for different probiotics and different wall materials. Finally, the spray drying microencapsulation technique becomes an efficient alternative to incorporate these microorganisms into different food matrices, making it an option for the food industry.

**Keywords:** atomization, microparticles, drying, probiotics

---

* Corresponding Author's Email: nataly.costa@ufv.br.

In: Microencapsulation of Probiotics
Editors: D. de Almeida Paula, K. Cristina Guedes Silva and N. de Almeida Costa
ISBN: 979-8-88697-979-4

## Introduction

The drying process consists of a widely used technique to promote food conservation, guarantee a stable and prolonged shelf life of products, reduce the size and consequently transport costs, and facilitate storage and marketing. Among the numerous drying techniques, spray drying is one of the most used in the food industry (Schuck et al. 2016; Martín et al. 2015; Gouin 2004; Rosenberg, Talmon, and Kopelman 1988).

Although, in most cases, spray drying was used to dehydrate and obtain different powder products, this technique can also be used to encapsulate active material such as probiotic microorganisms within a protective matrix formed by polymers (Castro-López et al. 2021; Martín et al. 2015; Dziezak 1988). Spray drying is one of the oldest techniques for microencapsulation of different compounds, its use began around 1930 to microencapsulate different flavors using acacia gum as wall material (Dziezak 1988; Shahidi, and Han 1993).

From the 1950s, this technique began to be used in the food industry to encapsulate aromas, provide greater protection, and prevent degradation or oxidation of the compound. Later, spray drying also started to be applied in the microencapsulation of bioactive molecules and probiotic microorganisms (Pu, Bankston, and Sathivel 2011; Gouin 2004; Ré 1998). The spray drying process was developed for the manufacture of powdered milk. However, milk drying can be considered as an encapsulation, as lactose and milk proteins function as a wall material capable of enveloping the milk fat, protecting it against oxidation (Gharsallaoui et al. 2007).

Currently, it is considered one of the oldest, most common, and most economical techniques used for the microencapsulation of different food ingredients (Furuta, and Neoh 2021). In addition, it avoids the risk of biological and chemical degradation of the core material, ensures greater stability due to low activity values and water content, and can be used on both laboratory and industrial scales (Rashidinejad et al. 2022; Gómez et al. 2018).

Spray drying consists of a unit operation where a liquid product is atomized into small droplets of about 10 to 150 μm through an atomizing nozzle in parallel flow to hot and dry air with a temperature above 100°C and, due to this an increase in the air-liquid interface area increases the drying kinetics and in a few seconds a powder is obtained (Gharsallaoui et al. 2007; Radosavljević et al. 2022).

The drying process of the water present in the solution occurs very quickly and then the probiotic microorganism is trapped. The choice of the wall material to be used will depend on the characteristics desired for the microparticles obtained, such as thermal resistance, resistance to acidic pH, rupture of the protective wall and release of the core, survival of the probiotics during passage through the gastrointestinal system, solubility to the be added in food matrices, storage stability and among others.

## Microencapsulation Process of Probiotics by Spray Drying

The microencapsulation technique by spray drying can transform liquid solutions, containing the wall material or combinations, together with the probiotic microorganism into a dry particulate powder. This process consists of 4 stages, according to Vivek et al. (2023).

Initially, even before the atomization of the liquid, the encapsulating solution, dispersion, or emulsion containing the compound to be protected must be prepared (Dziezak 1988; Shahidi and Han 1993). The principle of microencapsulation by spray drying consists of enveloping the probiotics with an external layer capable of protecting these microorganisms while feeding the equipment with the encapsulating solution containing the chosen wall material, or a combination of different materials (Gómez et al. 2018).

The first stage of the microencapsulation process by spray drying consists of atomizing this encapsulating solution containing the probiotic microorganism into small droplets by an atomizing nozzle that works through pressure or centrifugal energy (Gharsallaoui et al. 2007). The droplet formation stage determines the size and distribution of the formed microspheres (Dalmoro, Barba, Lamberti, and D'Amore 2012). Different atomizers can be used in this process, including a pneumatic atomizer, pressure nozzle, rotating disc, two fluid nozzles, and sonic nozzle (Masters 1968; Bowen 1938).

In atomization, there is the formation of an aerosol or suspension of small droplets composed of wall material and probiotics, in a gaseous phase usually formed by air at high temperature inside the drying chamber (Oxley 2014), as described in Figure 1.

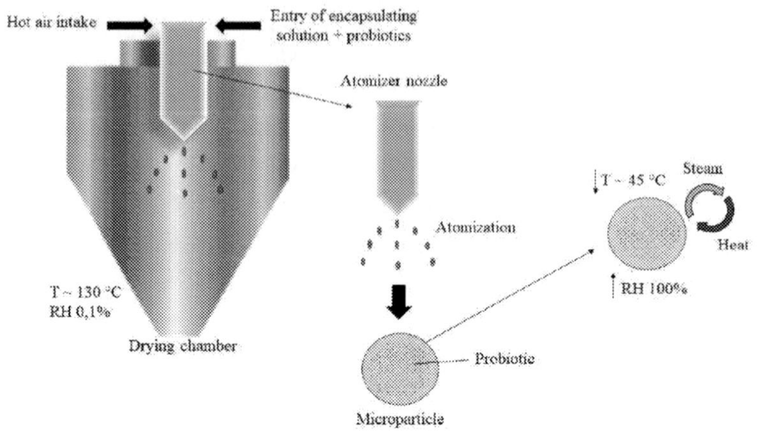


**Figure 1.** Atomization of the encapsulating solution containing probiotic microorganisms.

The main objective of the atomization step of the encapsulating solution is to create a maximum heat transfer surface between the dry air present in the drying chamber and the liquid (Figure 1). In the second stage of microencapsulation of probiotics, the atomized droplets come into contact with hot air, of approximately 120 to 150°C, supplied to the drying chamber.

According to Vivek et al. 2023, the microencapsulation of probiotics by spray drying can be divided into 4 phases. In the first phase, the encapsulating solution containing the probiotics is atomized through the atomizing nozzle, forming small droplets. In the second phase, the droplets formed come into contact with hot air supplied to the drying chamber. Subsequently, in the third phase, the drying of the droplets occurs, obtaining the microparticles. Finally, in the fourth phase, the separation of solid particles occurs in the drying chamber.

When the droplets formed to come into contact with the hot air present in the drying chamber, there is an equilibrium between the temperature and the partial pressure of the vapor between the liquid phase (droplet) and the gaseous phase (air). Thus, there is a transfer of heat between the air and the droplet due to differences in temperature and relative humidity.

The higher the energy supplied, the finer the droplets formed and for the same amount of energy, the size of the formed particles increases according to the increase in the feed rate, the viscosity of the encapsulating solution, and its surface tension (Gharsallaoui et al. 2007).

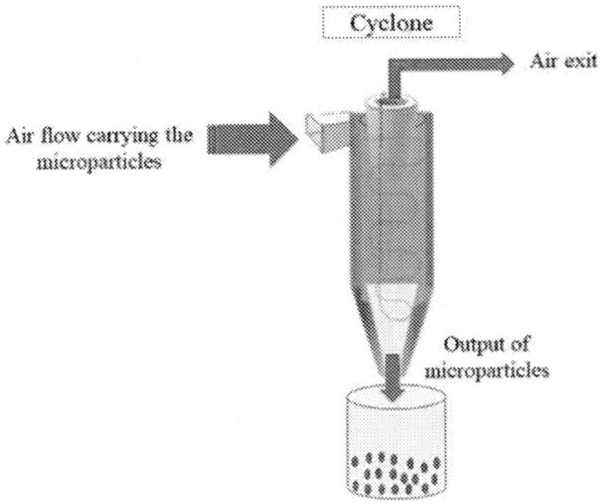


**Figure 2.** Representation of the cyclone for the separation of microparticles.

In the third stage, the atomized droplets are dried, and the powder is obtained, which are the microparticles containing the probiotic microorganisms. Due to its properties, the wall material dries at a much faster rate than the core material and this allows the wall material to form a coating around the droplet containing the probiotics (Gómez et al. 2018). As the exposure time of the microparticles to high temperatures is short, usually a few seconds, the temperature in the core of the microparticles generally remains between 50 to 80°C, which minimizes thermal damage to the probiotic (Azeredo 2008; Desai and Park 2005).

In the fourth step, the microparticles formed through drying are carried by the hot air flow toward the cyclone, where they are separated (Figure 2).

The cyclone separator is connected to the drying chamber to recover the powder particles dehydrated after the process (Gouin 2004; Mahdavi et al. 2014). The denser particles are recovered at the base of the drying chamber while the finer ones pass through the cyclone to be separated from the humid air (Gharsallaoui et al. 2007). This separation that occurs in the cyclone, in addition to reducing the losses of the product obtained in the atmosphere, also reduces the exposure of microparticles to the high temperature of the drying chamber.

In addition to the cyclones, the spray dryer is also equipped with some filters, called "bag houses". As the airflow manages to carry the formed particles, the larger ones are deposited in the cyclone and later collected. However, the finer and lighter particles follow the airflow and to prevent these particles from being spread in the ambient air, these filters are used to remove this finer dust (Estevinho, Rocha, Santos, and Alves 2013). The microencapsulation of probiotics by spray drying, the process steps, and the parts of the equipment were represented in Figure 3.

## Characteristics of the Obtained Microparticles

The microparticles obtained must protect the probiotic microorganism against adverse conditions such as pH variations and temperature variations. In addition, microparticles must release the probiotic at the site of action and in adequate amounts to promote benefits (Corona-Hernandez et al. 2013; Anekella and Orsat 2013).

The microparticles obtained by the spray drying technique are normally of the matrix type, that is, the probiotic microorganisms remain distributed in the encapsulating agent. Thus, the formed microparticles do not have a defined nucleus and the probiotic release mechanism occurs by diffusion (Azeredo 2005). The size of the microparticles obtained will depend on the operating conditions of the equipment, the type and size of the atomizing nozzle, and the wall material used for microencapsulation, and may vary from 5 to 1000 μm (Kreuter 1996; Liu, Chen, and Selomulya 2015). The sprayed droplets must be relatively small in size, to optimize drying and form smaller microparticles with small variations in size (Parikh 2005).

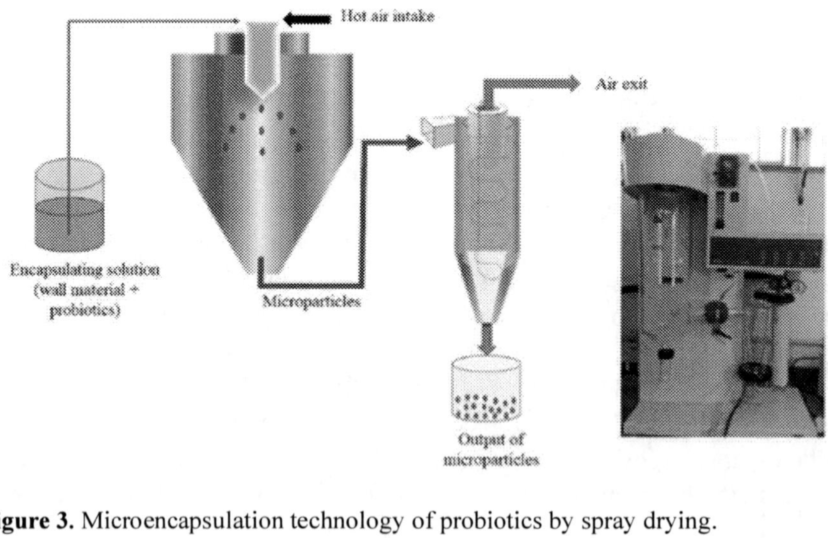


**Figure 3.** Microencapsulation technology of probiotics by spray drying.

The microencapsulation of probiotics can be performed in different ways and the microparticles obtained have distinct characteristics that depend on the combination of wall materials used, as well as the concentration of each compound. Such characteristics will influence the physical, chemical, and physical-chemical nature of the obtained microparticles (Jafari et al. 2022).

The composition of the surface that constitutes the microparticles plays a fundamental role in the characteristics and properties of the powdered microparticles obtained at the end of the process. Therefore, knowing the formation process and the characteristics of the materials that make up the wall of microparticles is essential to improve the microencapsulation process by spray drying (Porras-Saavedra et al. 2015).

Some characteristics of the microparticles developed, such as the water absorption capacity, is a critical factor that directly influences the stability of the microparticles produced, because when the powders receive moisture, there is a change in the physical structure of the microparticles, including the collapse of the structure and release of the material. encapsulation, stickiness, and loss of specific functions (Bhandari and Howes 1999).

## Advantages and Disadvantages of the Technique

Microencapsulation by spray drying is a relatively low-cost technology, a fast and continuous flow process, reproducible and scalable when compared to other techniques used for microencapsulation of sensitive compounds and, therefore, the most used at an industrial level (Pu, Bankston, and Sathivel 2011; Schafroth et al. 2012). Food manufacturers containing probiotic microorganisms use spray drying microencapsulation more frequently due to its simplicity, ease of operation, high processing efficiency, and controllable microcapsule size (Farahmandi et al. 2021).

Another important factor is that microencapsulation by spray drying allows the use of different types of wall materials with easy adaptation, contributing to several coating possibilities (Wilkowska et al. 2016). However, the encapsulating agent must be soluble in water since most spray-drying processes in the food industry are carried out from aqueous formulations (Desai and Park 2005).

The microparticles obtained in the form of a dry powder can extend the useful life of the coated compound, in addition, it is a material that is easy to store due to its compact volume (Xie et al. 2023). In addition, the microparticles produced by spray drying when added to the food matrix cause few changes to the characteristics of the product, especially in texture (Wilkowska et al. 2016; Jafari et al. 2022). As the powder obtained is very fine, and when it presents good solubility (> 75%), its dispersion in the matrix is fast without causing changes in the characteristics of the product.

On the other hand, the spray drying method is generally carried out at a temperature of around 50°C to 220°C. The temperature used in the drying process is an important factor that directly influences the survival of probiotics, as low inlet temperatures when used can reduce the efficiency of water evaporation, causing the particulate matter to accumulate, and high temperatures damage cell membranes, reducing the ability of probiotics to survive (Farahmandi et al. 2021).

Thus, some disadvantages of the spray drying microencapsulation technique are related to the temperature used for drying and obtaining the microparticles. In general, the damage caused to probiotics during the process is mainly influenced by the outlet temperature and not the inlet temperature (Arepally and Goswami 2019). Thermal, osmotic, and oxidative stress are the main mechanisms that cause probiotic inactivation during and after spray drying (Santivarangkna, Kulozik, and Foerst 2008). The reduction in the viability of the probiotic microorganism occurs due to dehydration and thermal inactivation (Perdana et al. 2014).

The high drying temperature causes the cellular pores to leak intracellular substances (Anekella and Orsat 2013). Furthermore, when the temperature of the microparticles reaches the temperature of the outlet drying air, the probiotic cells may suffer some thermal injuries capable of affecting the integrity of the membrane, causing denaturation of some proteins and damage to DNA, RNA, and ribosomes (Huang et al. 2017). The removal of water present in the hydration layer of probiotic cells directly affects the integrity of the cell membrane, causing different changes in its functions, or even in the structure of some enzymes and proteins present in the cell membrane.

Some alternatives can be used to reduce the mortality of probiotic microorganisms during drying, such as the addition of protectors such as trehalose, maltodextrin, and gum acacia to the encapsulating solution before drying (Anal and Singh 2007; Rashidinejad et al. 2022). The selection of the strain used and the choice of wall materials also contribute to greater encapsulation yield, greater encapsulation efficiency, and greater resistance to thermal stress, some compounds considered prebiotics are being added to the coating matrix of the developed microparticles (Peredo et al. 2016; Fiocco et al. 2020; Raddatz et al. 2020; Yuan et al. 2023). Prebiotics are considered substrates used selectively by host microorganisms conferring health benefits (Gibson et al. 2017).

Fritzen-Freire et al. (2012) evaluated the viability and physical properties of microcapsules containing *Bifidobacterium* BB-12 obtained by spray drying with partial replacement of the wall material (reconstituted skimmed milk - RSM) by prebiotic ingredients such as inulin, oligofructose and inulin enriched with oligofructose. The authors found that all microcapsules produced showed a high survival rate of microorganisms during storage for 180 days at temperatures of 4°C and -18°C. However, microcapsules that were produced with inulin and oligofructose-enriched inulin had higher initial counts. However, the combination of prebiotics such as inulin and oligofructose with the conventional wall material (reconstituted skimmed milk - RSM) resulted in better protection from bifidobacteria during storage. In addition, the authors verified that the addition of prebiotics to the wall material promoted greater thermal stability to the microcapsules produced and, consequently, greater viability of the microorganism.

Another factor that also influences the viability of the microencapsulated probiotic is the submission of these microorganisms to a stress factor before the drying process. Triggering adaptation responses to bacterial stress before drying is a strategy capable of improving viability during exposure to high temperatures during drying and during long storage, where thermal, oxidative, and osmotic adaptations are closely related to survival after spray drying (Huang et al. 2017).

Bommasamudram, Muthu, Devappa (2022) evaluated the thermal adaptation potential of the probiotic *Lacticaseibacillus casei* about thermal stress and spray drying. In this work, wall materials such as maltodextrin, corn starch, and gum Arabic were used. The probiotic microorganism was subjected to sublethal heat stress at 45 and 50°C for 60 min before spray drying. After exposure to high temperatures in spray drying, the authors observed that survival was increased when compared to treatment without exposure to sub-lethal stress. Furthermore, probiotic powders under sublethal heat stress showed better survivability when stored for 90 days at 4°C. In this case, the cells revealed overexpression of heat shock proteins, indicating a change in protein structure. Thus, the authors suggest sublethal heat stress as an alternative that can be used before spray drying to increase the viability of probiotics.

## Application of the Technique

Different research have been carried out on the microencapsulation of probiotics by spray drying using different strains and different wall materials, some studies are described in Table 1.

**Table 1.** Research on probiotic microencapsulation using spray drying

| Drying parameters | Microorganism | Wall material | References |
|---|---|---|---|
| Atomizer nozzle: 0.7mm<br>Inlet: 130°C<br>Output: 70°C<br>Flow: 25 mL/min | *Lacticaseibacillus casei* ATCC 335 | Symbiotic drinks based on legumes, maltodextrin, acacia gum, and xanthan gum | (Chaturvedi and Chakraborty 2022). |
| Inlet: 170°C<br>Output: 80°C | *Lactiplantibacillus plantarum* 21805 | Whey Protein Isolate and Dextran Conjugates | (Guo et al. 2022) |
| Flow rate: 110 mL/min<br>Inlet: 200°C<br>Output: 90°C | *Bifidobacterium animalis* ssp. *Lactis* ATCC27536 e *Lactobacillus acidophilus* ATCC4356 | Whey protein hydrolyzate and D-10 maltodextrin | (Minj and Anand 2022). |
| Flow: 7 mL/min<br>Inlet: 130 and 150°C<br>Output: not informed | *Limosilactobacillus reuteri* KUB-AC5 e *Lactobacillus johnsonii* KUNN19-2 | Whey Protein, Sucrose, and Whey Protein Isolate | (Sompach et al. 2022). |
| Inlet: 130°C, 140°C and 150°C<br>Output: 55 ± 3°C | *Lactobacillus acidophilus* NCDC 016 | Maltodextrin and Gum Arabic | (Arepally, Reddy, and Goswami 2020). |
| Inlet: 160°C<br>Output: 60 ± 1°C | *Lactobacillus rhamnosus* HN001 | Arabic gum, agave fructans, maltodextrin, inulin and trehalose | (Barajas-Alvarez, Gonzalez-Avila, and Espinosa-Andrews 2022) |
| Input: 150 ± 3°C<br>Output: not informed<br>Flow: 8.33 mL/min | *Lacticaseibacillus rhamnosus* GG | Whey Protein Isolate and Shellac | (Yin et al. 2022) |
| Atomizer nozzle: 0.5mm<br>Input: 110 ± 2°C<br>Output: 55 ± 3°C<br>Flow rate: 4 mL/min | *Lactobacillus plantarum* MTCC 5422 | Fructooligosaccharide (FOS), whey protein, and denatured whey protein isolate | (Rajam and Anandharama krishnan 2015) |

| Drying parameters | Microorganism | Wall material | References |
|---|---|---|---|
| Entry: 90°C<br>Output: 75°C<br>Supply flow: 0.25 L/h−1 | *Lactobacillusus plantarum* ATCC 8014 | Whey, whey permeate or retentate | (Eckert et al. 2017) |
| Entry: 170°C<br>Output: not informed<br>Supply flow: 8 mL/min | *L. rhamnosus* PTCC 1637 | Skimmed milk powder, gum Arabic, sucrose, and gum tragacanth. | (Farahmandi et al. 2021) |
| Entry: 120°C<br>Output: 65 – 70°C<br>Supply flow: 8 mL/min | *Limosilactobacillus fermentum* NCU001464 | Maillard reaction products of gelatin and xylooligosaccharides. | (Li et al. 2023) |
| Entry: 160°C<br>Output: 64 – 68°C<br>Supply flow: 9 mL/min | *Lactobacillus casei Shirota* | Maltodextrin reconstituted skimmed milk and gum Arabic. | (Gul 2017) |
| Entry: 195°C<br>Output: 85°C | *Lactobacillus acidophilus* LA-5, *Bifidobacterium animalis subsp. lactis* BB-12 *e Propionibacterium jensenii* 702 | Reconstituted goat's milk. | (Ranadheera et al. 2015). |

As can be seen, different wall materials can be used in the microencapsulation of probiotics by spray drying. Among the most used materials, acacia gum, maltodextrins, modified starch and different combinations of these compounds stand out. Some other proteins like whey protein and soy protein, sodium caseinate, alginate, carboxymethylcellulose, and guar gum are also commonly applied as encapsulating agents.

In addition to the works mentioned above in Table 1, other forms of spray drying microencapsulation are also commonly used. Zavaleta et al. (2022) evaluated the co-microencapsulation of *Lacticaseibacillus rhamnosus* in double emulsion with krill oil followed by spray drying using whey protein as wall material. In this study, inlet air temperature of 170 ± 5 or 180 ± 5°C and outlet air temperature of 90 ± 5°C were used. At both evaluated drying temperatures, *L. rhamnosus* showed viability above 10 log CFU.ml-1. In this study, it was possible to observe that the drying temperature had no significant effect on the viability of *L. rhamnosus*. The co-microencapsulation of probiotics has several advantages related to cost and efficiency and has been widely used to protect sensitive compounds (Misra, Pandey, and Mishra 2021; Vaziri, Alemzadeh, and Vossoughi 2019).

Arslan-Tontul and Erbas (2017) evaluated the microencapsulation of *Saccharomyces boulardii*, *Lactobacillus acidophilus*, and *Bifidobacterium bifidum* by spray drying combined with spray drying in single and double layers. This technique is very similar to spray drying, however, in this case, the atomization is carried out in a cold chamber. Thus, the main mechanism of the spray chilling technique is the injection of an encapsulating matrix into the cold chamber and the rapid solidification of the drops when in contact with cold air. The droplets then become microcapsules containing probiotics (Chambi et al. 2008). In this study, both techniques were used and as wall material gum arabic and β-cyclodextrin were used for the spray drying technique and hydrogenated palm oil was used for the chilling spray technique. The authors verified that the microorganisms *S. boulardii*, *L. acidophilus*, and *B. bifidum* presented high survival in the microparticles obtained by a single layer produced by spray chilling. In addition, double-layer microencapsulation contributed to greater thermal and gastric resistance of *L. acidophilus*. As evidenced by this study, there is a need to increase the viability of probiotics during production, digestion, and storage and this can be achieved by combining microencapsulation techniques.

Oliveira et al. (2007) evaluated the microencapsulation of *Bifidobacterium lactis* and *Lactobacillus acidophilus* using the complex coacervation technique with casein and pectin combined with the spray drying technique. The authors found that microencapsulated probiotics showed greater resistance to acidic conditions than the free probiotic. Therefore, the process used to combine the two microencapsulation techniques and the wall materials used was efficient in protecting the microorganisms under study.

Muhammad et al. (2017) evaluated different wall materials such as potassium alginate, pectin, and resistant potato starch combined with thermal protectors such as maltodextrin, whey protein isolate (WPI), and D-mannose, to offer greater protection to the probiotic *Lactobacillus plantarum* KLDS 1.0344. The combination of wall materials with heat-protecting agents was efficient to promote the survival of probiotics during the storage period. The formulations tested, the microparticles using resistant potato starch with thermo protector (maltodextrin and D-mannose) showed effective results to minimize the effects of thermal denaturation and improve the survival of probiotics against storage and passage through the gastrointestinal system simulated *in vitro*.

## Conclusion

As for the development and application of the spray drying microencapsulation technique, there are still some challenges to be overcome, among them the greater application of organic wall materials, to incorporate the concept of green wall materials or "green label" to produce microparticles. resistant to the protection of probiotic microorganisms.

The progress of research using microencapsulation of probiotics by spray drying is constant and the search for new encapsulation wall materials with some important characteristics such as coming from natural sources, with high availability and low cost. In addition, these materials must guarantee protection and high survival for probiotic microorganisms when subjected to adverse conditions. Simultaneously, the new materials must ensure the controlled release of the probiotic at the appropriate site of action.

For a wide application of the spray drying microencapsulation technique, the food and pharmaceutical industries still face some challenges, since the process of growth and drying of microorganisms occurs in a specific way for each probiotic strain. In addition, most of the research carried out on probiotic microencapsulation using spray drying was carried out on a laboratory scale, in which case, the expansion of this research to industrial conditions using high-capacity dryers is necessary. These research are fundamental for the expansion of the technique at an industrial level and should be approached in future works to evaluate the production and application of probiotic microparticles obtained and applied at an industrial level in different food products.

## References

Anal, Anil Kumar, and Singh, Harjinder. (2007). "Recent advances in microencapsulation of probiotics for industrial applications and targeted delivery." *Trends in food science & technology*, 18 (5), 240-51. https://doi.org/10.1016/j.tifs.2007.01.004

Anekella, Kartheek, and Orsat, Valérie. (2013). "Optimization of microencapsulation of probiotics in raspberry juice by spray drying." *LWT-Food Science and Technology*, 50 (1), 17-24. https://doi.org/10.1016/j.lwt.2012.08.003

Arepally, Divyasree, and Goswami, Tridib Kumar. (2019). Effect of inlet air temperature and gum Arabic concentration on encapsulation of probiotics by spray drying. *LWT-Food Science and Technology*, 99, 583-93. https://doi.org/10.1016/ j.lwt.2018.10.022

Arepally, Divyasree, Reddy, Ravula Sudharshan, and Goswami, Tridib Kumar. (2020). "Encapsulation of *Lactobacillus acidophilus* NCDC 016 cells by spray drying:

characterization, survival after *in vitro* digestion, and storage stability." *Food & function*, 11, 8694-706. https://doi.org/10.1039/D0FO01394C.

Arslan-Tontul, Sultan, and Erbas, Mustafa. (2017). "Single and double layered microencapsulation of probiotics by spray drying and spray chilling." *LWT-Food Science and Technology*, 81, 160-69. https://doi.org/10.1016/ j.lwt.2017.03.060.

Azeredo, Henriette Monteiro Cordeiro. (2005). "Encapsulação: aplicação à tecnologia de alimentos." *Alimentos e Nutrição Araraquara* ["Encapsulation: application to food technology." *Araraquara Food and Nutrition*], 16 (1), 89-97.

Barajas-Alvarez, Paloma, Gonzalez-Avila, Marisela, and Espinosa-Andrews, Hugo. (2022). "Microencapsulation of *Lactobacillus rhamnosus* HN001 by spray drying and its evaluation under gastrointestinal and storage conditions." *LWT-Food Science and Technology*, 153, 112485. https://doi.org/10.1016/j.lwt.2021.112485

Bhandari, B. R., and Howes, T. (1999). "Implication of glass transition for the drying and stability of dried foods." *Journal of Food Engineering*, 40 (1-2), 71-79. https://doi.org/10.1016/S0260-8774(99)00039-4.

Bommasamudram, Jyothna, Muthu, Arjun, and Devappa, Somashekar. (2022). "Effect of sub-lethal heat stress on viability of *Lacticaseibacillus casei* N in spray-dried powders." *LWT-Food Science and Technology*, 155, 112904. https://doi.org/10.1016/j.lwt.2021.112904.

Bowen, W. S. (1938). "Spray drying." *Industrial & Engineering Chemistry*, 30(9), 1001-002.

Castro-López, Cecilia, Espinoza-González, Carlos, Ramos-González, Rodolfo, Boone-Villa, V. Daniel, Aguilar-González, Miguel A., Martínez-Ávila, Guillermo C., Aguilar, Cristóbal N., and Ventura-Sobrevilla, Janeth M. (2021). "Spray-drying encapsulation of microwave-assisted extracted polyphenols from Moringa oleifera: Influence of tragacanth, locust bean, and carboxymethyl-cellulose formulations." *Food Research International*, 144, 110291. https://doi.org/10.1016/j.foodres.2021.110291.

Chambi, H. N. M., Alvim, I. D., Barrera-Arellano, D., and Grosso, C. R. F. (2008). "Solid lipid microparticles containing water-soluble compounds of different molecular mass: Production, characterization, and release profiles." *Food Research International*, 41(3), 229-36. https://doi.org/10.1016/j.foodres.2007.11.012.

Chaturvedi, Smriti, and Chakraborty, Snehasis. (2022). "Comparative analysis of spray-drying microencapsulation of *Lacticaseibacillus casei* in synbiotic legume-based beverages." *Food Bioscience*, 50, 102139. https://doi.org/10.1016/j.fbio.2022.102139.

Corona-Hernandez, Rócio I., Álvarez-Parrilla, Emilio, Lizardi-Mendoza, Jaime, Islas-Rubio, Alma R., De La Rosa, Laura A., and Wall-Medrano, Abraham. (2013). "Structural stability and viability of microencapsulated probiotic bacteria: a review." *Comprehensive Reviews in Food Science and Food Safety*, 12(6), 614-28. https://doi.org/10.1111/1541-4337.12030.

Dalmoro, Annalisa, Barba, Anna Angela, Lamberti, Gaetano, and d'Amore, Matteo. (2012). "Intensifying the microencapsulation process: Ultrasonic atomization as an innovative approach." *European Journal of Pharmaceutics and Biopharmaceutics*, 80(3), 471-77. https://doi.org/10.1016/j.ejpb.2012.01.006

Desai, Kashappa Goud. H., and Jin Park, Hyun. (2005). "Recent developments in microencapsulation of food ingredients." *Drying Technology*, 23(7), 1361-394. https://doi.org/10.1081/DRT-200063478.

Dziezak, J. D. (1988). "Microencapsulation and encapsulated ingredients." *Food technology (Chicago)*, 42(4), 136-53.

Eckert, Camila, Serpa, Vanessa Garcia, dos Santos, Adriani Cristina Felipe, da Costa, Simone Marinês, Dalpubel, Viviane, Lehn, Daniel Neutzling, and de Souza, Claucia Fernanda Volken. (2017). "Microencapsulation of *Lactobacillus plantarum* ATCC 8014 through spray drying and using dairy whey as wall materials." *LWT-food science and technology*, 82, 176-83. https://doi.org/10.1016/j.lwt.2017.04.045.

Estevinho, Berta Nogueiro, Rocha, Fernando, Santos, Lúcia, and Alves, Arminda. (2013). "Microencapsulation with chitosan by spray drying for industry applications–A review." *Trends in food science & technology*, 31(2), 138-55. https://doi.org/10.1016/j.tifs.2013.04.001.

Farahmandi, Kajal, Rajab, Shadi, Tabandeh, Fatemeh, Shahraky, Mahvash Khodabandeh, Maghsoudi, Amir, and Ashengroph, Morahem. (2021). "Efficient spray-drying of *Lactobacillus rhamnosus* PTCC 1637 using total CFU yield as the deciding factor." *Food Bioscience*, 40, 100816. https://doi.org/10.1016/j.fbio.2020.100816.

Fiocco, Daniela, Longo, Angela, Arena, Mattia Pia, Russo, Pasquele, Spano, Giuseppe, and Capozzi, Vittorio. (2020). "How probiotics face food stress: They get by with a little help." *Critical reviews in food science and nutrition*, 60(9), 1552-580. https://doi.org/10.1080/10408398.2019.1580673.

Fritzen-Freire, Carlise B., Prudêncio, Elane S., Amboni, Renata D. M. C., Pinto, Stephanie S., Negrão-Murakami, Aureanna. N., and Murakami, Fabio S. (2012). "Microencapsulation of bifidobacteria by spray drying in the presence of prebiotics." *Food Research International*, 45(1), 306-12. https://doi.org/10.1016/j.foodres.2011.09.020.

Furuta, Takeshi, and Neoh, Tze Loon. (2021). "Microencapsulation of food bioactive components by spray drying: A review." *Drying Technology*, 39 (12), 1800-831. https://doi.org/10.1080/07373937.2020.1862181.

Gharsallaoui, Adem, Roudaut, Gaelle, Chambin, Odile, Voilley, Andrée, and Saurel, Rémi. (2007). "Applications of spray-drying in microencapsulation of food ingredients: An overview." *Food research international*, 40(9), 1107-121. https://doi.org/10.1016/j.foodres.2007.07.004.

Gómez, Belén, Barba, Francisco J., Domínguez, Rubén, Putnik, Predag, Kovačević, Danijela Bursac, Pateiro, Mirian, Toldrá, Fidel, and Lorenzo, Jose M. (2018). "Microencapsulation of antioxidant compounds through innovative technologies and its specific application in meat processing." *Trends in Food Science & Technology*, 82, 135-47. https://doi.org/10.1016/j.tifs.2018.10.006.

Gouin, Sébastien. (2004). "Microencapsulation: an industrial appraisal of existing technologies and trends." *Trends in food science & technology*, 15(7-8), 330-47. https://doi.org/10.1016/j.tifs.2003.10.005.

Gul, Osman. (2017). "Microencapsulation of *Lactobacillus casei* Shirota by spray drying using different combinations of wall materials and application for probiotic dairy dessert." *Journal of Food Processing and Preservation*, 41(5), e13198. https://doi.org/10.1111/jfpp.13198.

Guo, Qi, Li, Shidong, Tang, Jiaxin, Chang, Shuaidan, Qiang, Liyue, Du, Gengan, Yue, Tianli, and Yuan, Yahong. (2022). "Microencapsulation of *Lactobacillus plantarum* by spray drying: Protective effects during simulated food processing, gastrointestinal conditions, and in kefir." *International Journal of Biological Macromolecules*, 194, 539-45. https://doi.org/10.1016/j.ijbiomac.2021.11.096.

Huang, Song, Vignolles, Marie-Laure, Chen, Xiao Dong, Le Loir, Yves, Jan, Gwénael, Schuck, Pierre, and Jeantet, Romain. (2017). "Spray drying of probiotics and other food-grade bacteria: A review." *Trends in food science & technology*, 63, 1-17. https://doi.org/10.1016/j.tifs.2017.02.007.

Jafari, Saeid, Jafari, Seid Mahdi, Ebrahimi, Mahdi, Kijpatanasilp, Isaya, and Assatarakul, Kitipong. (2022). "A decade overview and prospect of spray drying encapsulation of bioactive from fruit products: Characterization, food application and *in vitro* gastrointestinal digestion." *Food Hydrocolloids*, 108068. https://doi.org/10.1016/j.foodhyd.2022.108068.

Kreuter, J. (1996). "Nanoparticles and microparticles for drug and vaccine delivery." *Journal of Anatomy*, 189(Pt 3), 503.

Li, Huiyu, Peng, Fei, Lin, Jia-Xin, Xiong, Tao, and Huang, Tao. (2023). "Preparation of probiotic microcapsules using gelatin-xylooligosaccharides conjugates by spray drying: Physicochemical properties, survival, digestion resistance, and colonization." *Food Bioscience*, 52, 102462. https://doi.org/10.1016/j.fbio.2023.102462.

Liu, W., Chen, X. D., and Selomulya, C. (2015). "On the spray drying of uniform functional microparticles." *Particuology*, 22, 1-12.https://doi.org/10.1016/j.partic.2015.04.001.

Mahdavi, Sahar Akhavan, Jafari, Seid Mahdi, Ghorbani, Mohammad, and Assadpoor, Elham. (2014). "Spray-drying microencapsulation of anthocyanins by natural biopolymers: A review." *Drying Technology*, 32(5), 509-18. https://doi.org/10.1080/07373937.2013.839562.

Martín, María José, Lara-Villoslada, Federico, Ruiz, María Adolfina, and Morales, María Encarnación. (2015). "Microencapsulation of bacteria: A review of different technologies and their impact on the probiotic effects." *Innovative Food Science & Emerging Technologies*, 27, 15-25. https://doi.org/10.1016/j.ifset.2014.09.010.

Masters, Keith. (1968). "Spray Drying—The Unit Operation Today." *Industrial & Engineering Chemistry*, 60(10), 53-63.

Minj, Shayanti, and Anand, Sanjeev. (2022). "Development of a spray-dried conjugated whey protein hydrolysate powder with entrapped probiotics." *Journal of Dairy Science*, 105(3), 2038-048. https://doi.org/10.3168/jds.2021-20978.

Misra, Sourav, Pandey, Pooja, and Mishra, Hari Niwas. (2021). "Novel approaches for co-encapsulation of probiotic bacteria with bioactive compounds, their health benefits and functional food product development: A review." *Trends in Food Science & Technology*, 109, 340-51https://doi.org/10.1016/j.tifs.2021.01.039

Muhammad, Zafarullah, Ramzan, Rabia, Huo, Gui-Cheng, Tian, Hui, and Bian, Xin. (2017). "Integration of polysaccharide-thermoprotectant formulations for microencapsulation of *Lactobacillus plantarum*, appraisal of survivability and physical-biochemical properties during storage of spray dried powders." *Food hydrocolloids*, 66, 286-295. https://doi.org/10.1016/j.foodhyd.2016.11.040.

Oliveira, Alessandra Cristina, Moretti, T. S., Boschini, C., Baliero, J. C. C., Freitas, O. D., and Fávaro-Trindade, C. S. (2007). "Stability of microencapsulated B. lactis (BI 01) and *L. acidophilus* (LAC 4) by complex coacervation followed by spray drying." *Journal of Microencapsulation*, 24(7), 685-93. https://doi.org/10.1080/02652040701532908.

Oxley, James. (2014). "Overview of microencapsulation process technologies." In *Microencapsulation in the food industry* (pp. 35-46). Academic Press. https://doi.org/10.1016/B978-0-12-404568-2.00004-2.

Parikh, D. M. (2005). Handbook of pharmaceutical granulation technology. *Drugs and the pharmaceutical sciences*, 81.

Perdana, Jimmy, Fox, Martijin B., Siwei, Chen, Boom, Remko M., and Schutyser, Maarten A. I. (2014). "Interactions between formulation and spray drying conditions related to the survival of *Lactobacillus plantarum* WCFS1." *Food Research International*, 56, 9-17. https://doi.org/10.1016/j.foodres.2013.12.007.

Peredo, A. G., Beristain, C. I., Pascual, L. A., Azuara, E., and Jimenez, M. (2016). "The effect of prebiotics on the viability of encapsulated probiotic bacteria." *LWT-Food Science and Technology*, 73, 191-96. https://doi.org/10.1016/j.lwt.2016.06.021.

Porras-Saavedra, J., Palacios-González, E., Lartundo-Rojas, L., Garibay-Febles, V., Yáñez-Fernández, J., Hernandez-Sanchez, H., Gutiérrez-López, G., and Alamilla-Beltran, L. (2015). "Microstructural properties and distribution of components in microparticles obtained by spray-drying." *Journal of food engineering*, 152, 105-12. https://doi.org/10.1016/j.jfoodeng.2014.11.014.

Pu, Jianing, Bankston, J. David, and Sathivel, Subramaniam. (2011). "Developing microencapsulated flaxseed oil containing shrimp (*Litopenaeus setiferus*) astaxanthin using a pilot scale spray dryer." *Biosystems engineering*, 108(2), 121-32. https://doi.org/10.1016/j.biosystemseng.2010.11.005.

Raddatz, Greice Carine, Poletto, Gabriela, de Deus, Cassandra, Codevilla, Cristiane Franco, Cichoski, Alexandre José, Jacob-Lopes, Eduardo, Muller, Edson Irineu, Flores, Erico Marlon Moraes, Esmerino, Erick Almeida, de Menezes, Cristiano Ragagnin. (2020). "Use of prebiotic sources to increase probiotic viability in pectin microparticles obtained by emulsification/internal gelation followed by freeze-drying." *Food Research International*, 130, 108902. https://doi.org/10.1016/j.foodres.2019.108902.

Radosavljević, Milos, Lević, Steva, Pejin, Jelena, Mojović, Ljiljana, and Nedović, Viktor. (2022). Encapsulation technology of lactic acid bacteria in food fermentation. In *Lactic Acid Bacteria in Food Biotechnology* (pp. 319-347). Elsevier.

Rajam, R., and Anandharamakrishnan, C. (2015). "Microencapsulation of *Lactobacillus plantarum* (MTCC 5422) with fructooligosaccharide as wall material by spray drying." *LWT-Food Science and Technology*, 60(2), 773-80. https://doi.org/10.1016/j.lwt.2014.09.062.

Ranadheera, C. Senaka, Evans, Craig A., Adams, Michelle. C., and Baines, Surinder K. (2015). "Microencapsulation of *Lactobacillus acidophilus* LA-5, Bifidobacterium animalis subsp. lactis BB-12 and *Propionibacterium jensenii* 702 by spray drying in goat's milk." *Small Ruminant Research*, 123(1), 155-59.https://doi.org/10.1016/j.smallrumres.2014.10.012.

Rashidinejad, Ali, Bahrami, Akbar, Rehman, Abdur, Rezaei, Atefe, Babazadeh, Afshin, Singh, Harjinder, and Jafari, Seid Mahdi. (2022). Co-encapsulation of probiotics with prebiotics and their application in functional/synbiotic dairy products. *Critical reviews in food science and nutrition*, 62(9), 2470-494. https://doi.org/10.1080/10408398.2020.1854169.

Ré, M. I. (1998). Microencapsulation by spray drying. *Drying Technology*, 16(6), 1195-236.

Rosenberg, M., Talmon, Y., and Kopelman, I. J. 1988. The microstructure of spray-dried microcapsules. *Food Structure*, 7(1), 14.

Santivarangkna, C., Kulozik, U., and Foerst, P. (2008). "Inactivation mechanisms of lactic acid starter cultures preserved by drying processes." *Journal of applied microbiology*, 105(1), 1-13. https://doi.org/10.1111/j.1365-2672.2008.03744.x.

Schafroth, Nina, Arpagaus, Cordin, Jadhav, Umesh. Y., Makne, Sushil, Douroumis, Dennis. (2012). "Nano and microparticle engineering of water-insoluble drugs using a novel spray-drying process." *Colloids and surfaces B: Biointerfaces*, 90(1), 8-15. https://doi.org/10.1016/j.colsurfb.2011.09.038.

Schuck, Pierre, Jeantet, Romain, Bhandari, Bhesh, Chen, Xiao Dong, Perrone, Ítalo Tuler, Carvalho, Antônio Fernandes, Fenelon, Mark, and Kelly, Phil. (2016). "Recent advances in spray drying relevant to the dairy industry: A comprehensive critical review." *Drying Technology*, 34(15), 1773-790. https://doi.org/10.1080/07373937.2016.1233114.

Shahidi, F., and Han, X. Q. (1993). "Encapsulation of food ingredients." *Critical Reviews in Food Science & Nutrition*, 33(6), 501-47. https://doi.org/10.1080/10408399309527645.

Sompach, Gessirin, Rodklongtan, Akkaratch, Nitisinprasert, Sunee, and Chitprasert, Pakamon. (2022). "Microencapsulating role of whey protein isolate and sucrose in protecting the cell membrane and enhancing the survival of probiotic lactobacilli strains during spray drying, storage, and simulated gastrointestinal passage." *Food Research International*, 159, 111651. https://doi.org/10.1016/j.foodres.2022.111651.

Vaziri, Asma Sadat, Alemzadeh, Iran, and Vossoughi, Manouchehr. (2019). "Survivability and oxidative stability of co-microencapsulated *L. Plantarum* PTCC 1058 and DHA as a juice carrier." *Food Bioscience*, 32, 100460. https://doi.org/10.1016/j.fbio.2019.100460.

Vivek, Kambhampati, Mishra, Sabyasachi, Pradhan, Rama Chandra, Nagarajan, Meena, Kumar, Pavitra Krishna, Singh, Singam Suranjoy, Manvi, Dronachari, and Gowda, Na Nanje. (2022). "A Comprehensive Review on Microencapsulation of Probiotics: Technology, Carriers and Current Trends." *Applied Food Research*, 100248. https://doi.org/10.1016/j.afres.2022.100248.

Wilkowska, Agnieszka, Ambroziak, Wojciech, Czyzowska, Agata, and Adamiec, Janusz. (2016). "Effect of microencapsulation by spray-drying and freeze-drying technique on the antioxidant properties of blueberry (*Vaccinium myrtillus*) juice polyphenolic compounds." *Polish Journal of Food and Nutrition Sciences*, 66(1), 11. https://doi.org/10.1515/pjfns-2015-0015.

Xie, Aijun, Zhao, Shanshan, Liu, Zifei, Yue, Xiqing, Shao, Junhua, Li, Mohan, and Li Zhiwei. (2023). "Polysaccharides, proteins, and their complex as microencapsulation carriers for delivery of probiotics: A review on carrier types and encapsulation techniques." *International Journal of Biological Macromolecules*, 124784. https://doi.org/10.1016/j.ijbiomac.2023.124784.

Yin, Ming, Yuan, Yongkai, Chen, Maoshen, Liu, Fei, Saqib, Md Nazmus, Chiou, Bor-Sen, and Zhong, Fang. (2022). "The dual effect of shellac on survival of spray-dried *Lactobacillus rhamnosus* GG microcapsules." *Food Chemistry*, 389, 132999. https://doi.org/10.1016/j.foodchem.2022.132999

Yonekura, Lina, Sun, Han, Soukoulis, Christos, and Fisk, Ian. (2014). "Microencapsulation of *Lactobacillus acidophilus* NCIMB 701748 in matrices containing soluble fibre by spray drying: Technological characterization, storage stability, and survival after *in vitro* digestion." *Journal of functional foods*, 6, 205-14. https://doi.org/10.1016/j.jff.2013.10.008.

Yuan, Cong, Hu, Ronghai, He, Li, Hu, Jielun, and Liu, Huan. (2023). "Extraction and prebiotic potential of β-glucan from highland barley and its application in probiotic microcapsules." *Food Hydrocolloids*, 139, 108520. https://doi.org/10.1016/j.foodhyd.2023.108520.

Zavaleta, E. Bonilla, Coavichi, L. I. Lozano, Rodríguez, L. C. Velasco, Andrade, E. Flores, García, Hugo S., and Díaz, M. Rascón. (2022). "Co-microencapsulation of *Lactobacillus rhamnosus* and krill oil by spray-drying." *Food Bioscience*, 50, 102133. https://doi.org/10.1016/j.fbio.2022.102133.

# Chapter 6

# Microencapsulation by Emulsification

**Daniele de Almeida Paula*, PhD**
Federal Institute of Paraná, IFPR, Campus Assis Chateaubriand, Brazil

## Abstract

Nowadays, the demand for probiotic foods is increasing, however, the industry faces challenges related to maintaining the viability of probiotic cells. Thus, the encapsulation process, which is based on protecting probiotic cells when subjected to adverse conditions, can be used. Among the microencapsulation techniques, the emulsification process stands out due to mild processing conditions, thus maintaining the integrity and viability of the probiotic cells. Water-in-oil (W/O) and water-in-oil-in-water (W/O/W) emulsions can be used. The incorporation of prebiotics can also be performed in the internal phase of emulsions associated with probiotic cells to improve cell viability. Encapsulation of probiotics using the emulsification process represents an advantageous and attractive method to increase the viability of living organisms. In this chapter, important advances in emulsification techniques are discussed, including the development of double emulsions.

**Keywords:** functional foods, probiotic microorganisms, emulsification

## Introduction

The Food and Agriculture Organization of the United Nations and World Health Organization (FAO/WHO) define probiotics as live microorganisms

* Corresponding Author's Email: daniele.paula@ifpr.edu.br.

In: Microencapsulation of Probiotics
Editors: D. de Almeida Paula, K. Cristina Guedes Silva and N. de Almeida Costa
ISBN: 979-8-88697-979-4

that, when administered in hich when consumed in adequate amounts, confer a health benefit to the consumer (FAO/WHO, 2001). Therefore, they are widely recognized for promoting health and well-being, probiotics have related beneficial effects, including the modulation of the intestinal microbiota, through the inhibition of pathogenic microorganisms, the production of anticancer compounds, immune system modulation, etc. (Rodrigues et al. 2020).

However, the beneficial effects provided by probiotics depend on the recommended regular and daily intake, to provide the minimum "dose" of viable cells. It is suggested that the number of cells in food containing probiotic bacteria should be $10^8$ - $10^9$ $CFUg^{-1}$ to ensure that, after ingestion, at least $10^6$ -$10^7$ $CFUg^{-1}$ can reach the specific point of action (small and/or large intestine) and exert the desired effect (Ahmad et al. 2018). Conditions during processing, inadequate transport methods, variation in storage temperature, and exposure to gastrointestinal fluids are among the factors that can decrease the number of viable cells.

Thus, microencapsulation is one of the most efficient strategies that have been considered in recent years to protect probiotic cells from adverse conditions by keeping them in sufficient amounts to exert their beneficial effects (Calabuig-Jiménez et al. 2019). At the same time, the global market for probiotic supplements and foods has been growing significantly, and cell encapsulation has emerged as an alternative to incorporate probiotics into different food matrices. (Y. Zhang et al. 2022).

Microencapsulation is defined as a system where active materials are coated with encapsulating materials. The formed capsule is often called the shell capsule, carrier material, matrix, or membrane. The encapsulation protects the probiotic cells against various physical-chemical stresses such as moisture, heat, pH, and harmful substances, thus increasing the survival (Vivek et al. 2023).

Currently, various encapsulation techniques are used. However, the encapsulation process using emulsification offers advantages in that it is possible to use gentle and non-aggressive operations, as well as non-toxic solvents during the process. Emulsification is a technique often used in food processing to encapsulate enzymes, vitamins, minerals, and probiotic microorganisms. This is due to the simplicity of choosing the necessary components for the formulation, preparation conditions, and direct procedures involved in the production.

The emulsion is a heterogeneous and thermodynamically unstable system, defined as the mixture of two immiscible liquids, where one is dispersed into the other in the form of droplets. For example, if oil is the dispersed phase the emulsion is named oil in water (O/W); if water is the dispersed phase the emulsion is known as water in oil (W/O). Systems more complex are denominated double emulsions (Flaiz et al. 2016) and the most common form is water in oil in water (W/O/W) and oil in water in oil (O/W/O) (Pawlik and Norton 2014).

Emulsions, such as water in oil (W/O) and water in oil in water (W/O/W), can be used to encapsulate and protect probiotics in gastrointestinal conditions. The encapsulation of probiotic microorganisms using the emulsification process is an efficient alternative to maintain both the viability and the stability of the encapsulated cells.

Thus, the proposed chapter will present the use of microencapsulation through the emulsification process as an alternative for maintaining the stability and survival of probiotic microorganisms during processing, storage, and passage through the gastrointestinal systems.

## Emulsions

The emulsion is a heterogeneous and thermodynamically unstable system, defined as the mixture of two immiscible liquids, where one is dispersed into the other in the form of droplets (Dickinson 2011). In the food industry products such as mayonnaise, soups, yogurt, sauces, cheese, ice cream, butter, and margarine are examples of simple emulsions.

Simple emulsions are classified according to the type of dispersed phase. For example, if oil is in the dispersed phase the emulsion is named oil in water (O/W); if water is in the dispersed phase the emulsion is known as water in oil (W/O). Systems more complex are denominated double emulsions (Flaiz et al. 2016) and the most common form is water in oil in water (W/O/W) and oil in water in oil (O/W/O) (Pawlik and Norton 2014). The types of emulsion are described in Figure 1.

Recent research and studies show the use of emulsions, such as water in oil (W/O) and water in oil in water (W/O/W) to encapsulate and protect probiotic cells from adverse conditions, such as gastrointestinal conditions. In addition, it appears that the dispersed aqueous phase (W/O or W/O/W emulsion) is preferentially used, due to the hydrophilic nature of bacterial cells. For example, Picone et al. (2017) described that encapsulation of probiotic cells in water-in-oil (W/O) emulsions increased cell viability after digestion under simulated *in vitro* conditions.

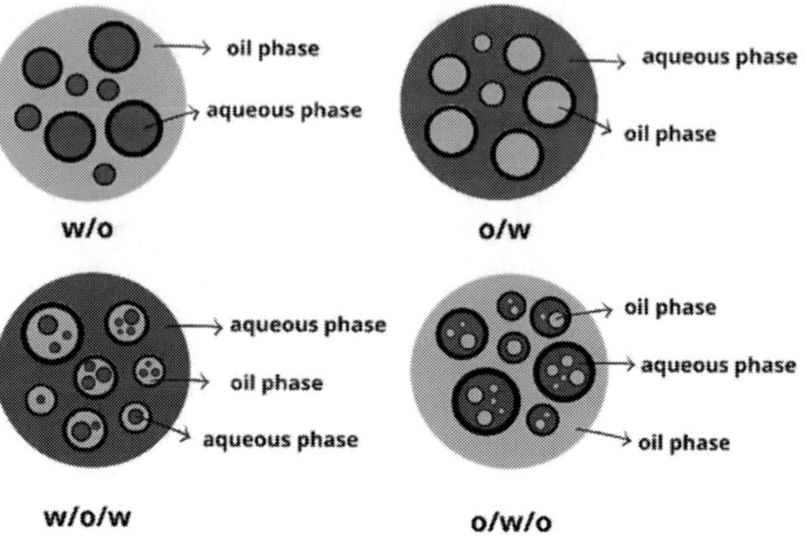


**Figure 1**. Simple emulsion (W/O and O/W) and double emulsion (W/O/W and O/W/O).

Qi, Luo, and Lu (2019) used W/O emulsions to produce microgels containing probiotic cells, improving cell viability after *in vitro* simulation of digestion. Wang et al. (2020) reported the encapsulation of *Lactobacillus acidophilus* AS 1.2686 using the double emulsification process. According to the results, the authors verified an increase in the viability of encapsulated probiotic cells during 14 days of storage compared to free cells. Under simulated gastrointestinal conditions, approximately 84% of encapsulated cells remained viable after digestion, while free cells presented a rapid decline and total loss of viability, principally due to poor tolerance to acidic pH (Rodrigues et al. 2020).

## Simple Emulsions

For the formation of emulsions, considerable participation of energy is required usually mechanic energy (Leal-Calderon, Thivilliers, and Schmitt 2007). As a result, devices capable of high shear or high pressure are used. For example, a high-speed or ultrasonic homogenizer can be used to prepare small particles, while a microfluidizer can be used to produce nano- and microemulsions. Often, a combination of physical and chemical methods is used to prepare these emulsions (Haji et al. 2022).

As previously mentioned, for the production of emulsions mechanical shear forces are normally employed aiming at breaking the dispersed phase into very small droplets, leading to a substantial increase in the interfacial area (Ho et al. 2022). The process is unstable thermodynamically thus the emulsions do not form spontaneously. Generally, a third component or combination of several surface-active agents named emulsifiers is used (Yamanaka et al. 2017). Various emulsifiers are used to kinetically stabilize food emulsions. Emulsifiers used for emulsion stability can be divided into the following four categories: low molecular weight emulsifiers, proteins, polysaccharides, and solid particles (Cai et al. 2023).

Emulsifiers are amphiphilic molecules that have both hydrophilic and lipophilic properties and are also named surfactants. The main role of emulsifiers is to adsorb at the interface between oil and water of newly formed droplets. Thus, there is a reduction in the interface tension and energy required for the formation of the emulsion. Furthermore, the formation of a film around the drops with steric and electrostatic properties can occur, avoiding unstable phenomena that cause the separation of the phases of the system. Accordingly, Dickinson (2011) the emulsifiers most utilized in the food industry are proteins, phospholipids, and in certain cases polysaccharides (hydrocolloids).

The proteins have the most surface activity among emulsifiers compared to polysaccharides or hydrocolloids, thus, high concentration can stabilize the emulsion (Dickinson 2011). Recently, the stabilization of emulsions by conjugates and hydrocolloid complexes with food proteins is a thriving research area. Soluble polysaccharides (pectin, alginate, gelana) have also been added in the external aqueous phase to act as stabilizing agents due to their thickening/gelling properties (Dickinson 2011). Choosing specific emulsifiers or a blend of emulsifiers for a determined function is critical to the creation of stable emulsions (Chung and McClements 2013).

## Encapsulation of Probiotics in Simple Emulsion

Cedran et al., (2022) evaluated the optimization of water-in-oil emulsions containing *Limosilactobacillus reuteri* DSM 20016, with various surfactant concentrations (5, 10, and 20 g/Kg) and W/O ratios (30:70, 40:60 and 50:50). According to the results obtained, the authors concluded that the two most stable emulsions had 40:60 W/O and 10 or 20 g/kg of surfactant. Both samples provided high efficiency of encapsulation and protection during storage and *in vitro* gastrointestinal tests.

Zhang et al., (2022) used polyglycerol polyricinoleate (PGPR) as the oil-soluble surfactant, and beeswax was used as the oil phase to formulate an emulsion (W/O HIPEs) for the encapsulation of probiotic cells. High internal phase water-in-oil emulsions (W/O HIPEs) consist of a high concentration of water droplets dispersed in a continuous oil phase. In the work carried out, the probiotic cells were in the internal aqueous phase of the W/O emulsions, surrounded by a layer of oil. After simulating *in vitro* gastrointestinal conditions, the authors verified that the encapsulation of probiotics increased their survival. This study provides useful information on the use of W/O to improve the effectiveness of probiotics in the food industry.

Figure 2, in summary form, shows how the microencapsulation of probiotic cells can be performed using simple emulsions.

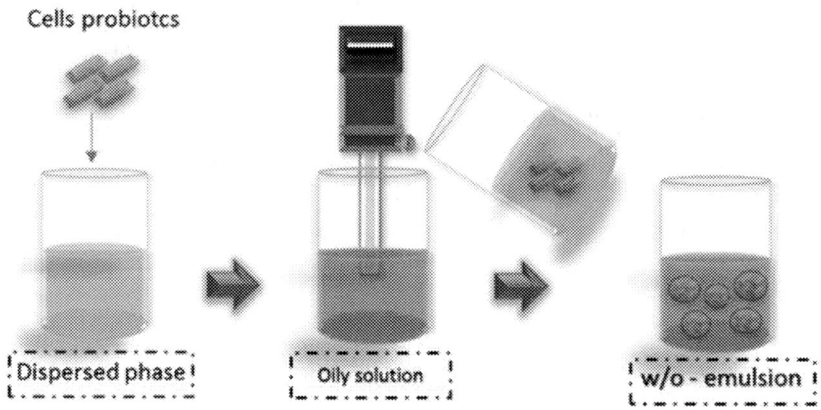


**Figure 2**. Encapsulation of probiotics in simple emulsion.

Elvan; Baysal; Harsa, (2022) evaluated the microencapsulation of *Lactiplantibacillus pentosus* and its impregnation to coat the surfaces of table olives. Microencapsulation using the water-in-oil emulsion and the polymers (whey protein concentrate and xylan) improved the viability of *L. pentosus*, which successfully impregnated the surface of olives. Furthermore, after storage at 4°C for four weeks, microencapsulated *L. pentosus* showed cell levels above the recommended limit for functional foods. In this way, according to the results obtained, the ingestion of about 10 to 15 olives per serving (corresponding to 40 to 60 g of olives) provides the recommended minimum daily dose, which varies between $10^7$ to $10^8$ UFC.

## Emulsification/Internal Ionic Gelation

A solidifying agent can be used to separate the dispersion phase droplets from the continuous phase (Zhang, Zhang, and McClements 2016). This procedure has often been suggested as an alternative to external extrusion/gelling in the encapsulation of various compounds.

In this technique, a hydrocolloid solution containing the probiotics cells and a non-ionized solidifying agent is added to the continuous oil phase of a water-in-oil (W/O) emulsion. Internal gelation of the dispersed droplets containing the hydrocolloid can be initiated by the release of $Ca^{2+}$ from an insoluble complex (calcium salt) by lowering the pH (Ahmed et al. 2013).

These methods show great promise as a tool for the development of encapsulation processes, especially for the new field of nanotechnology using natural polymers (Reis et al. 2006).

Raddatz et al. (2020) used mixtures based on pectin, inulin, and rice bran to encapsulate *Lactobacillus acidophilus* LA-5. The obtained particles protected the probiotic cells in simulated gastrointestinal conditions and improved the encapsulation efficiency, reaching rates above 90%. Furthermore, the bacteria remained viable during storage for 120 days at 25°C.

## Emulsification/External Ionic Gelation

The properties that differentiate external and internal gelation are pore size, rigidity, permeability, and matrix strength. Alginate microspheres prepared with external gelling have smaller pores and a denser surface structure than those inside the capsule. In the external gelling emulsification method, an alginate solution containing the encapsulated active substance is emulsified in the oil phase to form a W/O emulsion. Then a cross-linking compound, e.g., calcium chloride solution, is gradually added to the emulsion. (Łętocha, Miastkowska, and Sikora 2022). Consequently, the calcium ions move from the surface of the drop to its interior (Schoubben et al. 2010).

Song et al. (2013) used the process of emulsification/external gelation and emulsification/internal gelation to produce microcapsules for the retention of probiotics. In the study, the properties of microencapsulated cells were evaluated, including morphology, size and size distribution, cell viability, characteristics of microorganisms, and survival rates under gastrointestinal conditions.

According to the results obtained, it was verified that there was no difference in cell viability during the preparation between the techniques. However, the morphology, size, and size distribution of microcapsules by internal emulsification/gelling were superior to external emulsification/gelling techniques. After *in vitro* simulation of gastrointestinal conditions, it was found that the emulsification/gelling process can be used for potential application in the food or biotechnology industries.

## Double Emulsions

Double emulsions are complex dispersion systems called "emulsions of emulsions" that exist simultaneously with both types of single emulsions (O/W) and (W/O) and due to their properties and structures, they are especially interesting. The most common forms are water in oil in water (W/O/W) and oil in water in oil (O/W/O) (Pawlik and Norton 2014), with the former being the most studied.

In the emulsion (W/O/W) there are simultaneously water droplets into oil globules (O) and dispersed in the continuous aqueous phase (W). Because they consist of three distinct phases and two oil-water interfaces, compared to the two phases and one water-oil interface of a simple emulsion, it is necessary to use two emulsifying agents with different values of the hydrophilic-lipophilic balance (HLB). In this case, an emulsifier is necessary to stabilize the droplets of the internal emulsion (main) and another to stabilizer the droplets of the external emulsion (secondary) (Eslami, Davarpanah, and Vahabzadeh 2017).

Among surfactants, polyglycerol polyricinoleate is widely used, with a BHL value equal to 1. PGPR is a synthetic surfactant also named polyglycerol ester of ricinoleic acid and represents the most hydrophobic emulsifier used in foods and is recognized as security by FDA (Food and Drug Administration) (McClements, 1994). Hydrophilic emulsifiers for the O/W interface are more diverse. Small synthetic surfactants such as Tween 80 and 60 can be used, however, polysaccharides and proteins are predominantly applied due to lower migration and ability to form viscoelastic layers (Øye et al. 2023). The emulsifier molecules already at the interface will undergo new reorganization processes, thus forming a viscoelastic film at the oil-water interface by intermolecular crosslinking. These viscoelastic interfacial films can prevent oil droplet aggregation through electrostatic repulsion and steric hindrance effects (Niu et al. 2023).

Due to the physical-chemical properties and structure of emulsions, studies have been carried out to reduce the fat and sodium content, improve the sensory properties of products, or control the release and protection of substances and microorganisms (including the encapsulation of probiotics), aiming to maintain its viability during production, storage and gastrointestinal digestion.

According to Qin, Luo, and Li (2021), double emulsions are particularly suitable for encapsulating hydrophilic bioactive substances (such as probiotics) because they can be incorporated into the internal aqueous phase and thus protect from the external environment. Furthermore, prebiotics such as fructooligosaccharides have also been incorporated into the inner aqueous phase of dual emulsions to improve the survival rate of probiotics.

## Fabrication of Double Emulsions

The systems of double emulsions usually are prepared using a method of two steps. The first step is based on the elaboration of primary emulsion W/O or O/W using a lipophilic or hydrophilic emulsifier, respectively. In the second step, the primary emulsion is emulsified again by the addition of another aqueous phase or oil phase containing a hydrophilic or lipophilic emulsifier, respectively, thus forming W/O/W or O/W/O emulsions. (Dickinson, 2011). The double emulsion formation process is illustrated in Figure 3.

In the first moment, high shear forces are used, through the addition of high energy (high pressure, ultra-turrax, or ultrasound) (Muschiolik 2007). Subsequently, the second step occurs under moderate shear conditions, avoiding instability of the system (Dickinson, 2011). Generally, a high agitation speed is used for the primary emulsification and a lower speed for the secondary one. The very high or very low shear rates drastically affect system stability, determining shear/shake time optimization.

The conventional methods used generate substantially polydisperse and structurally heterogeneous systems. In this way, new low-shear and low-energy methods have been proposed and used. One can mention, among them, emulsification microchannels, microcapillary devices, and membrane emulsification. A disadvantage of using these methods is the low production rate when compared to traditional emulsification methods (Leal-Calderon, Thivilliers, and Schmitt 2007).

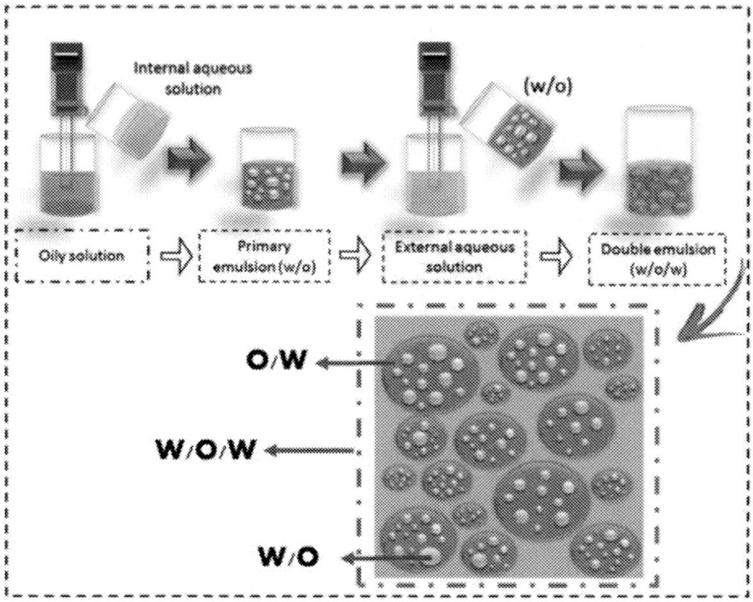


**Figure 3**. Fabrication of double emulsion.

## Encapsulation of Probiotics in Double Emulsion

Pimentel-González et al., (2009) evaluated the viability of probiotic cells of *Lactobacillus rhamnosus* included in the internal aqueous phase of a double emulsion of water in oil in water (W/O/W) in relation of cells in the external aqueous phase of the double emulsion (control), when exposed to conditions of low pH and in the presence of bile salts. Survival of trapped cells was significantly increased under conditions of low pH and bile salt, and their survival was 108% and 128%, respectively. The researchers concluded that the internal phase of the double emulsions provided the preparations with an appropriate environment for growth, increasing probiotic viability compared to the viability of cells contained in the external aqueous phase of the double emulsion.

García-Almend; Pimentel-Gonz; Lude, (2014) evaluated the inclusion of *Lactobacillus plantarum* cells in the internal aqueous phase in double emulsions (O/W/O) and addition during the manufacture of Oaxaca cheese. The results obtained indicated that the inclusion of *L. plantarum* in double emulsions protected the probiotic cells under the most diverse processing conditions of Oaxaca cheese and from the gastrointestinal conditions, reaching the colonization sites in numbers high enough to provide benefits to the health of their hosts.

Flores-Andrade et al., (2017) studied the impregnation of probiotic cells of *L. rhamnosus* included in the internal aqueous phase of a double emulsion using the methods of osmotic dehydration (OD) and vacuum osmotic dehydration (VOD) and evaluated the mass transfer during the impregnation in apple slices "Granny Smith" variety. As a result, the authors concluded that the double emulsions protect the microorganisms inside the apple cells and both the OD and VOD processes produced apple slices with probiotics content above $1.64 \times 10^7$ $CFU.g^{-1}$ corresponding to those found in commercial probiotic products.

Paula et al. (2019) when microencapsulating probiotic cells of *Lactobacillus plantarum* through a process consisting of double emulsification followed by complex coacervation, verified that after in vitro simulation of gastrointestinal conditions, the viability of encapsulated cells was 80.4%, while for free cells it was 25%. Furthermore, probiotic cell viability was maintained during storage at 8°C and -18°C for 45 days.

Lisboa et al. (2022) elaborated a mango jam formulation containing probiotic *L. plantarum* cells included in the internal phase of double emulsions and studied cell viability under simulated *in vitro* gastrointestinal conditions. The double emulsion protected the cell probiotics from extreme stress in the gastric phase, providing an appropriate growth environment. Thus, the inclusion of probiotic cells in non-dairy products, such as mango desserts, is of great relevance, meeting a consumer demand restricted to dairy products and increasing the diversity of probiotic products on the market.

El, Mantzouridou, and Gkatzionis (2018) evaluated the use of probiotic cells from *Lactobacillus paracasei subsp.* paracasei DC 412 incorporated into the W/O/W emulsion system in yogurt production. For this, the effect on fermentation, bacterial growth kinetics, physical-chemical properties, and structural characteristics was evaluated. The authors verified that during fermentation, the starter culture was not affected by the introduction of *L. paracasei* and W/O/W emulsion. Cell viability of encapsulated *L. paracasei* increased during storage and after exposure to simulated gastrointestinal conditions. These results demonstrate the ability of the W/O/W emulsion to be used for probiotic yogurt fortification to increase functionality without interfering with starter culture and fermentation.

## Characterization of Emulsions Used in the Microencapsulation of Probiotic Cells

### Viscosity and Rheology

Rheology is described as the science that evaluates how materials deform and flow under the influence of external forces (Pal 1996). Viscosity is the main rheological property of any emulsified system and is primarily affected by shear rate. Emulsions with a low concentration of dispersed phase (<0.60) exhibit Newtonian fluid behavior (Ilyasoglu Buyukkestelli and El 2019).

For Newtonian fluids, the viscosity remains constant even under variable strain rate conditions, called absolute viscosity. There is a linear relationship between shear stress and strain rate (Steffe, 1996). For those considered non-Newtonian, the flow curves (shear stress x strain rate) are not linear, that is, the viscosity values change with the variation in the strain rate values according to the characteristics of each fluid (Rao, 1999). With this, the property is called apparent viscosity ($\eta a$) (Pelegrine et al., 2000).

Several mathematical models help in the rheological characterization, the choice of which one to use depends on the information and the characteristic of the fluid. The most common models are Newton's, Bingham's or ideal plastic, Ostwald de Waele's or power law, and Herschel-Bulkley's or modified power (Machado, 2002). For the analysis of the rheological behavior, it is advisable to wait a few hours (24 – 48 h) since a freshly prepared emulsion takes a certain amount of time to reach the corresponding viscosity (Lachman; Liederman; Kanig 2001). The most used instruments for characterizing the rheological properties are shear devices (shear viscometers and dynamic shear rheometers)

### Droplet Size Distribution

Some process variables, such as agitation speed, phase addition order, and temperature, influence the final physicochemical characteristics of the emulsion. The different droplet sizes can be obtained depending on the chosen emulsification method (Moreira de Morais et al. 2006), changing the physical stability of these systems (e.g., gravitational separation, flocculation, coalescence, and Ostwald maturation) and their optical properties (e.g., brightness and color). Most emulsions are polydisperse systems that can be characterized as "monomodal," "bimodal" or "multimodal," depending on whether there are one, two, or more peaks in the particle size distribution.

The concentration of droplets within different size classes can be measured using a variety of commercially available instruments (McClements, 2005), for example, optical microscopy, confocal laser scanning microscopy, electron microscopy, static light scattering, dynamic light scattering, electrical pulse counting, sedimentation techniques, ultrasonic spectrometry, nuclear magnetic resonance, neutron scattering, dielectric spectroscopy (Clausse et al. 2005).

One of the most used techniques is dynamic light scattering which is based on the detection of the intensity of a beam of light that, when falling on the droplet, can be deflected, absorbed, or transmitted. For droplets larger than the wavelength of the emitted light, reflection, refraction or diffraction occurs, whereas for smaller ones, light scattering occurs (Jones 1999). Strategically placed detectors measure the intensity and angle of scattered light. Utilizing mathematical algorithms, the signal from the detectors is then converted to the droplet size distribution (Allen, 1997). Any system is capable of scattering light, but the greatest interest is in the study of dispersed or suspended molecules and pure systems. With light scattering, small or aggregated molecules are observed mainly in solutions, colloid dispersions, suspensions, and gels (Chu, 1991).

## Microscopy

The microscope is one of the main instruments for size analysis and for a complete characterization of the particle, such as its shape, morphology (surface) and color. Optical microscopes (reflected or transmitted light) can be used in image formation (Papini, 2003) and are widely used for analyzing the distribution of particles in the medium (Oliveira, 2007).

There are different techniques to study the microstructure of single and double emulsions, such as Confocal Laser Scanning Microscopy (CLSM). This technique is based on the incidence of light that causes fluorescence in the material, and obtaining three-dimensional images with appropriate software (Groves, 2006), representing an important advance in optical microscopy. Its limitation is related to the wavelengths of the equipment's laser, allowing or not a greater visualization of the sample.

## Potential Zeta

The zeta potential is a measure of the magnitude of electrostatic repulsion or attraction between charged particles, and it is one of the parameters that affect stability. Its measurement provides a detailed understanding of the causes of dispersion, aggregation, or flocculation and can be applied to improve the formulation of dispersions, emulsions, and suspensions (Malvern Ltd. 2011).

For the measurement of zeta potential, a hundred-fold dilution of freshly prepared emulsions is required. The zeta potential is highly affected by pH and the type of emulsifier present in the system (Kumar et al. 2022). Most particulate materials in contact with a liquid acquire an electrical charge on their surface, making them electrically charged.

This charge can appear in various ways, through the dissociation of ionizable groups and the differential adsorption of ions from the solution on the particle surface, increasing the concentration of counterions and forming an electric double layer at the liquid interface. This double layer is divided into two regions: the first (internal) region consists of ions strongly bound to the surface, while the other (external) region has a distribution determined by the balance between electrostatic forces and thermal movement (Schmickler 2014).

The value of the electric potential in the shear plane is called Zeta Potential (Bamba et al. 2018). Measuring the zeta potential, it is possible to delineate conditions of greater stability for emulsions Negative values of zeta potential result in repulsive interaction between particles, preventing aggregation and providing good physical and biochemical stability to the system.

The most common and accepted technique for measurement is through electrokinetic mobility. Firstly, the diluted colloidal suspension is introduced into a cell with two electrodes, and an electric potential is applied. The electrically charged species will move toward the electrode of the opposite charge, and the velocity of the particles depends on the surface charge and the intensity of the applied electric field (Malvern Ltd. 2011).

If all the species in suspension have a high (in absolute value) zeta potential, they will tend to repel each other. However, if they have a low (in absolute value) zeta potential, and in the absence of other mechanisms (steric hindrance, the high viscosity of the external phase, among others) that can prevent the aggregation of two or more adjacent particles, the system will destabilize (Réfega, 2010). Therefore, zeta potential, combined with other stabilization mechanisms, can be used as an indicator of the kinetic stability of a colloidal system, and the results are expressed in millivolts (mV).

## Conclusions and Prospects

Probiotic microorganisms are often encapsulated using spray drying or lyophilization techniques. However, the extreme temperatures and mechanical forces to which probiotic cells are subjected during these processes can result in a loss of cell viability. The emulsification encapsulation process can be implemented under mild processing conditions, thereby maintaining the integrity and viability of the probiotic cells.

The encapsulation of probiotics using the emulsification process represents an advantageous and appealing method for enhancing the viability of living organisms. In this chapter, important advancements in emulsification techniques are addressed, including the development of double emulsions.

## References

Ahmad, Bilal, Asir Gani, Adil Gani, Asima Shah, and Farooq Ahmad. 2018. "Production of RS4 from Rice Starch and Its Utilization as an Encapsulating Agent for Targeted Delivery of Probiotics." *Food Chemistry* 239: 287–94. https://doi.org/10.1016/j.foodchem.2017.06.110

Ahmed, Mahmoud M., Saleh Abd El-Rasoul, Sayed H. Auda, and Mohamed A. Ibrahim. 2013. "Emulsification/internal Gelation as a Method for Preparation of Diclofenac Sodium-Sodium Alginate Microparticles." *Saudi Pharmaceutical Journal* 21 (1): 61–69. https://doi.org/10.1016/j.jsps.2011.08.004

Allen, T. Particle Size Measurement, 5 * edition, v.l, London: Chapman and Hall, 1997.

Bamba, Bio Sigui Bruno, John Shi, Carole C. Tranchant, Sophia Jun Xue, Charles F. Forney, Loong Tak Lim, Weili Xu, and Guihua Xu. 2018. "Coencapsulation of Polyphenols and Anthocyanins from Blueberry Pomace by Double Emulsion Stabilized by Whey Proteins: Effect of Homogenization Parameters." *Molecules* 23 (10). https://doi.org/10.3390/molecules23102525

Cai, Zhixiang, Yue Wei, Aimin Shi, Jian Zhong, Pingfan Rao, Qiang Wang, and Hongbin Zhang. 2023. "Correlation between Interfacial Layer Properties and Physical Stability of Food Emulsions: Current Trends, Challenges, Strategies, and Further

Perspectives." *Advances in Colloid and Interface Science* 313 (February): 102863. https://doi.org/10.1016/j.cis.2023.102863

Calabuig-jiménez, Laura, Ester Betoret, Noelia Betoret, Francesca Patrignani, Cristina Barrera, Lucía Seguí, Rosalba Lanciotti, and Marco Dalla. 2019. "High Pressures Homogenization (HPH) to Microencapsulate *L. Salivarius Spp*. Salivarius in Mandarin Juice. Probiotic Survival and in Vitro Digestion." *Journal of Food Engineering* 240 (July 2018): 43–48. https://doi.org/10.1016/j.jfoodeng.2018.07.012

Cedran, M. F., F. J. Rodrigues, H. H. Sato, and J. L. Bicas. 2022. "Optimization of a Water-in-Oil Emulsion Containing *Limosilactobacillus Reuteri*: Applicability of Pequi Oil as a Continuous Phase." *Current Research in Biotechnology* 4 (April): 318–25. https://doi.org/10.1016/j.crbiot.2022.07.003

Chu, B. Laser light scattering – Basic principles and practice. 2. ed. San Diego: Academic Press Inc., 1991

Chung, Cheryl, and David Julian McClements. 2013 Structure–Function Relationships in Food Emulsions: Improving Food Quality and Sensory Perception." *Food Structure*, December, 1–21. https://doi.org/10.1016/j.foostr.2013.11.002

Clausse, D., F. Gomez, I. Pezron, L. Komunjer, and C. Dalmazzone. 2005. "Morphology Characterization of Emulsions by Differential Scanning Calorimetry." *Advances in Colloid and Interface Science* 117 (1–3): 59–74. https://doi.org/10.1016/j.cis.2005.06.003

Dickinson, Eric. 2011. "Double Emulsions Stabilized by Food Biopolymers." *Food Biophysics* 6 (1): 1–11. https://doi.org/10.1007/s11483-010-9188-6

El, Hani, Fanith Mantzouridou, and Konstantinos Gkatzionis. 2018. "Utilisation of Water-in-Oil-Water (W/O/W) Double Emulsion in a Set-Type Yogurt Model for the Delivery of Probiotic *Lactobacillus Paracasei*" 107 (February): 325–36. https://doi.org/10.1016/j.foodres.2018.02.049

Elvan, Menşure, Ayşe Handan Baysal, and Sebnem Harsa. 2022. "Microencapsulation of a Potential Probiotic *Lactiplantibacillus Pentosus* and Its Impregnation onto Table Olives." LWT 156 (September 2021): 0–7. https://doi.org/10.1016/j.lwt.2021.112975

Eslami, Parisa, Leila Davarpanah, and Farzaneh Vahabzadeh. 2017. "Encapsulating Role of Β -Cyclodextrin in Formation of Pickering Water- in-Oil-in-Water (W/O/W) Double Emulsions Containing *Lactobacillus Dellbrueckii.*" *Food Hydrocolloids* 64: 133–48. https://doi.org/10.1016/j.foodhyd.2016.10.035

Flaiz, Linda, María Freire, Susana Cofrades, Raquel Mateos, Jochen Weiss, Francisco Jimenez-Colmenero, and Ricard Bou. 2016. "Comparison of Simple, Double, and Gelled Double Emulsions as Hydroxytyrosol and N-3 Fatty Acid Delivery Systems." *Food Chemistry* 213: 49–57. https://doi.org/10.1016/j.foodchem.2016.06.005

Flores-Andrade, E., L. A. Pascual-Pineda, F. G. Alarcón-Elvira, M. P. Rascón-Díaz, D. J. Pimentel-Gonzalez, and C. I. Beristain. 2017. "Effect of Vacuum on the Impregnation of Lactobacillus Rhamnosus Microcapsules in Apple Slices Using Double Emulsion." *Journal of Food Engineering* 202: 18–24. https://doi.org/10.1016/j.jfoodeng.2017.02.005

García-almend, B E, D J Pimentel-gonz, and F Lude. 2014. "Viability of *Lactobacillus Plantarum* Entrapped in Double Emulsion during Oaxaca Cheese Manufacture, Melting, and Simulated Intestinal Conditions". "LWT - Food Science and Technology 59: 768–73. https://doi.org/10.1016/j.lwt.2014.07.004

Groves, K. Microscopy: A Toll to Study Ingredient Interactions in Foods. In: GAONKAR, A. G.; MCPHERSON, A. (Eds.). Ingredient Interactions - Effects on Food Quality. 2°. ed. Boca Raton: Taylor & Francis Gruop, 2006. p. 21-48

Haji, Fatemah, James Cheon, Jiyoo Baek, Qi Wang, and Kam Chiu Tam. 2022. "Application of Pickering Emulsions in Probiotic Encapsulation- A Review." *Current Research in Food Science* 5 (September): 1603–15. https://doi.org/10.1016/j.crfs.2022.09.013

Ho, Thao Minh, Aysan Razzaghi, Arun Ramachandran, and Kirsi S. Mikkonen. 2022. "Emulsion Characterization via Microfluidic Devices: A Review on Interfacial Tension and Stability to Coalescence." *Advances in Colloid and Interface Science* 299 (September 2021): 102541. https://doi.org/10.1016/j.cis.2021.102541

Ilyasoglu Buyukkestelli, H., and Sedef Nehir El. 2019. "Development and Characterization of Double Emulsion to Encapsulate Iron." *Journal of Food Engineering* 263 (April): 446–53. https://doi.org/10.1016/j.jfoodeng.2019.07.026

Jato, J. Aspectos fundamentales de los sistemas farmacéuticos y operações básicas - [Fundamental aspects of pharmaceutical systems and basic operations] Madrid, Editorial Sintesis, 1997.

Jones, A. R. 1999. "Light Scattering for Particle Characterization." *Progress in Energy and Combustion Science* 25 (1): 1–53. https://doi.org/10.1016/S0360-1285(98)00017-3

Kumar, Ankit, Ramandeep Kaur, Vikas Kumar, Satish Kumar, Rakesh Gehlot, and Poonam Aggarwal. 2022. "New Insights into Water-in-Oil-in-Water (W/O/W) Double Emulsions: Properties, Fabrication, Instability Mechanism, and Food Applications." *Trends in Food Science and Technology*. Elsevier Ltd. https://doi.org/10.1016/j.tifs.2022.07.016

Lachman, L.; Liederman, H. A.; Kanig, J. L. The theory and practice of industrial pharmacy, Lisboa, Fundação Calouste Gulbenkian, 2001

Leal-Calderon, F, F Thivilliers, and V Schmitt. 2007. "Structured Emulsions" 12: 206–12. https://doi.org/10.1016/j.cocis.2007.07.003

Łętocha, Anna, Małgorzata Miastkowska, and Elżbieta Sikora. 2022. "Preparation and Characteristics of Alginate Microparticles for Food, Pharmaceutical and Cosmetic Applications." *Polymers* 14 (18). https://doi.org/10.3390/polym14183834

Machado, J. C. V. Reologia e Escoamento de Fluidos - Ênfase na Indústria do Petróleo, ed. Interciência - [Rheology and Flow of Fluids - Emphasis in the Petroleum Industry, ed. interscience] 2002

McCLements, D.J. Food emulsions: Principles, Practice, and Techniques. 2. ed. CRC Press, 1994.

McClements, D.J. Food emulsions: Principles, practices, and techniques. CRC Press, Boca Raton, FL, USA, 2005.

Malvern Ltd. 2011. "Zeta Potential: An Introduction in 30 Minutes." *Zetasizer Nano Serles Technical Note. MRK654-01* 2: 1–6.

http://scholar.google.com/scholar?hl=en&btnG=Search&q=intitle:Zeta+Potential+An+Introduction+in+30+Minutes#0

Moreira de Morais, Jacqueline, Orlando David Henrique dos Santos, Tatiani Delicato, Ricardo Azzini Gonçalves, and Pedro Alves da Rocha-Filho. 2006. "Physicochemical Characterization of Canola Oil/Water Nano-emulsions Obtained by Determination of Required HLB Number and Emulsion Phase Inversion Methods." *Journal of Dispersion Science and Technology* 27 (1): 109–15. https://doi.org/10.1081/DIS-200066829

Muschiolik, Gerald. 2007. "Multiple Emulsions for Food Use." *Current Opinion in Colloid & Interface Science* 12 (4–5): 213–20. https://doi.org/10.1016/j.cocis.2007.07.006

Niu, Hui, Wenduo Wang, Zuman Dou, Xianwei Chen, Xianxiang Chen, Haiming Chen, and Xiong Fu. 2023. "Multiscale Combined Techniques for Evaluating Emulsion Stability: A Critical Review." *Advances in Colloid and Interface Science* 311 (October 2022): 102813. https://doi.org/10.1016/j.cis.2022.102813

Øye, Gisle, Sébastien Simon, Turid Rustad, and Kristofer Paso. 2023. "Trends in Food Emulsion Technology: Pickering, Nano-, and Double Emulsions." *Current Opinion in Food Science* 50: 101003. https://doi.org/10.1016/j.cofs.2023.101003

Pal, Rajinder. 1996. "Effect of Droplet Size on the Rheology of Emulsions" 42 (11): 3181–90.

Paula, Daniele de Almeida, Eliane Maurício Furtado Martins, Nataly de Almeida Costa, Patrícia Martins de Oliveira, Eduardo Basílio de Oliveira, and Afonso Mota Ramos. 2019. "Use of Gelatin and Gum Arabic for Microencapsulation of Probiotic Cells from *Lactobacillus Plantarum* by a Dual Process Combining Double Emulsification Followed by Complex Coacervation." *International Journal of Biological Macromolecules* 133: 722–31. https://doi.org/10.1016/j.ijbiomac.2019.04.110

Pawlik, Aleksandra K., and Ian T. Norton. 2014. "Bridging Benchtop Research and Industrial Processed Foods: Structuring of Model Food Emulsions." *Food Structure* 1 (1): 24–38. https://doi.org/10.1016/j.foostr.2013.10.002

Papini, C.J. Comparative study of methods for determining the size of particle. São Paulo. Master's Dissertation - Institute of Energy and Nuclear Research - University of São Paulo. 130p, 2003.

Pelegrini, D. H.; Vidal, J. R. M. B. Gasparetto, C. A. Study of aparent viscosity of mango (Keitt) and pineapple (Pérola) pulps. 2000. Food Sci. Technol 20 (1)

Picone, Carolina Siqueira Franco, Andressa Catozzi Bueno, Mariano Michelon, and Rosiane Lopes Cunha. 2017. "Development of a Probiotic Delivery System Based on Gelation of Water-in-Oil Emulsions." *LWT- Food Science and Technology* 86: 62–68. https://doi.org/10.1016/j.lwt.2017.07.045

Pimentel-gonzález, D J, R G Campos-montiel, C Lobato-calleros, and R Pedroza-islas. 2009. "Encapsulation of *Lactobacillus Rhamnosus* in Double Emulsions Formulated with Sweet Whey as Emulsifier and Survival in Simulated Gastrointestinal Conditions." *Food Research International* 42 (2): 292–97. https://doi.org/10.1016/j.foodres.2008.12.002

Oliveira, M. J. Microscopia Óptica. In: CANEVAROLO-JR, S. V. (Ed.). Técnicas de Caracterização de Polímeros. São Paulo, SP: Artliber Editora LTDA. p. 147-164, 2007

Qi, Liang, Zhigang Luo, and Xuanxuan Lu. 2019. "Modulation of Starch Nanoparticle Surface Characteristics for the Facile Construction of Recyclable Pickering Interfacial Enzymatic Catalysis." *Green Chemistry* 21 (9): 2412–27. https://doi.org/10.1039/c9gc00779b

Qin, Xin Sheng, Zhi Gang Luo, and Xiao Lin Li. 2021. "An Enhanced pH-Sensitive Carrier Based on Alginate-Ca-EDTA in a Set-Type W/O/W Double Emulsion Model Stabilized with WPI-EGCG Covalent Conjugates for Probiotics Colon-Targeted Release." *Food Hydrocolloids* 113 (June): 106460. https://doi.org/10.1016/j.foodhyd.2020.106460

Rao, M. Anandha; Rheology of Fluid and Semisolids Foods-Principles and Applications, Maryland, USA, Aspen Publication, 1999

Raddatz, Greice Carine, Bruna de Souza da Fonseca, Gabriela Poletto, Eduardo Jacob-Lopes, Alexandre José Cichoski, Edson Irineu Muller, Erico Marlon Moraes Flores, Cristiane de Bona da Silva, and Cristiano Ragagnin de Menezes. 2020. "Influence of the Prebiotics Hi-Maize, Inulin, and Rice Bran on the Viability of Pectin Microparticles Containing Lactobacillus Acidophilus LA-5 Obtained by Internal Gelation/emulsification." *Powder Technology* 362: 409–15. https://doi.org/10.1016/j.powtec.2019.11.114

Reis, Catarina P., Ronald J. Neufeld, Sandra Vilela, António J. Ribeiro, and Francisco Veiga. 2006. "Review and Current Status of Emulsion/dispersion Technology Using an Internal Gelation Process for the Design of Alginate Particles." *Journal of Microencapsulation* 23 (3): 245–57. https://doi.org/10.1080/02652040500286086

Réfega, R. J. M. Nanopartículas para Aplicação Biomédica. Dissertação de Mestrado - Instituto Superior Técnico – Universidade Técnica de Lisboa. p. 71, 2010

Rodrigues, F. J., M. F. Cedran, J. L. Bicas, and H. H. Sato. 2020b. "Encapsulated Probiotic Cells: Relevant Techniques, Natural Sources as Encapsulating Materials and Food Applications – A Narrative Review." *Food Research International* 137 (September). https://doi.org/10.1016/j.foodres.2020.109682

Schmickler, W. 2014. *Electrochemical Theory: Double Layer*. *Reference Module in Chemistry, Molecular Sciences, and Chemical Engineering*. Elsevier Inc. https://doi.org/10.1016/b978-0-12-409547-2.11149-7

Schoubben, Aurélie, Paolo Blasi, Stefano Giovagnoli, Carlo Rossi, and Maurizio Ricci. 2010. "Development of a Scalable Procedure for Fine Calcium Alginate Particle Preparation." *Chemical Engineering Journal* 160 (1): 363–69. https://doi.org/10.1016/j.cej.2010.02.062

Song, Huiyi, Weiting Yu, Meng Gao, Xiudong Liu, and Xiaojun Ma. 2013. "Microencapsulated Probiotics Using Emulsification Technique Coupled with Internal or External Gelation Process." *Carbohydrate Polymers* 96 (1): 181–89. https://doi.org/10.1016/j.carbpol.2013.03.068

Vivek, Kambhampati, Sabyasachi Mishra, Rama Chandra Pradhan, Meena Nagarajan, Pavitra Krishna Kumar, Singam Suranjoy Singh, Dronachari Manvi, and NA Nanje Gowda. 2023. "A Comprehensive Review on Microencapsulation of Probiotics: Technology, Carriers and Current Trends." *Applied Food Research* 3 (1): 100248. https://doi.org/10.1016/j.afres.2022.100248

Wang, Ling, Mingyue Song, Zijun Zhao, Xia Chen, Jiyang Cai, Yong Cao, and Jie Xiao. 2020. "*Lactobacillus Acidophilus* Loaded Pickering Double Emulsion with Enhanced Viability and Colon-Adhesion Efficiency." *LWT - Food Science and Technology* 121 (December 2019): 108928. https://doi.org/10.1016/j.lwt.2019.108928

Yamanaka, Yohei, Isao Kobayashi, Marcos A Neves, Sosaku Ichikawa, Kunihiko Uemura, and Mitsutoshi Nakajima. 2017. "LWT - Food Science and Technology Formulation of W/O/W Emulsions Loaded with Short-Chain Fatty Acid and Their Stability Improvement by Layer-by-Layer Deposition Using Dietary Fi Bers." *LWT - Food Science and Technology* 76: 344–50. https://doi.org/10.1016/j.lwt.2016.07.063

Zhang, Yun, Youfa Xie, Hang Liu, David Julian McClements, Ce Cheng, Liqiang Zou, Wenjun Liu, and Wei Liu. 2022. "Probiotic Encapsulation in Water-in-Oil High Internal Phase Emulsions: Enhancement of Viability under Food and Gastrointestinal Conditions." *Lwt* 163 (January): 113499. https://doi.org/10.1016/j.lwt.2022.113499

Zhang, Zipei, Ruojie Zhang, and David Julian McClements. 2016. "Encapsulation of β-Carotene in Alginate-Based Hydrogel Beads: Impact on Physicochemical Stability and Bioaccessibility." *Food Hydrocolloids* 61: 1–10. https://doi.org/10.1016/j.foodhyd.2016.04.036

# Chapter 7

# Complex Coacervation for the Microencapsulation of Probiotics

**Daniele de Almeida Paula*, PhD**
Federal Institute of Paraná, IFPR, Campus Assis Chateaubriand, Paraná, Brazil

## Abstract

Currently, several studies and techniques have been carried out to maintain the viability of probiotic cells. The probiotic food market is increasingly expanding, and consumers are also looking for new options. With this, the improvement and development of new products is necessary. However, the food industry faces several challenges regarding cell viability during the processing, storage, and consumption of probiotic foods. Consequently, several techniques can be employed for the microencapsulation of probiotics. Among the techniques, complex coacervation has been widely researched and studied. The complex coacervation process that occurs is based on the electrostatic interaction between oppositely charged molecules at a specific pH. Due to the interaction of the polymers, complex coacervates are formed. In the encapsulation process, these coacervates aggregate around the probiotics cells, forming a barrier and protecting them against adverse conditions. Complex coacervation is an emerging microencapsulation technique for probiotic cells considered advantageous. Thus, the presented chapter focuses on the encapsulation process of probiotic cells by complex coacervation, in addition to discussing the advantages and disadvantages of this process.

**Keywords:** functional foods, coacervation, biopolymers

---

* Corresponding Author's Email: daniele.paula@ifpr.edu.br.

In: Microencapsulation of Probiotics
Editors: D. de Almeida Paula, K. Cristina Guedes Silva and N. de Almeida Costa
ISBN: 979-8-88697-979-4

## Introduction

In 1965, the research published by Lilly and Stillwell (1964) used the term "probiotics" to describe "substances secreted by one microorganism stimulating the growth of another" (Song et al., 2023). Nowadays, due to advancements in research and technology, several studies demonstrate the beneficial health effects associated with the consumption of probiotics. According to the Food and Agriculture Organization of the United Nations (FAO) and the World Health Organization (WHO), probiotics are defined as "live microorganisms that, when consumed in adequate amounts, confer a health benefit on the host." The market for functional foods, including probiotics, has been growing significantly. However, the incorporation of probiotic cells into food still faces certain challenges. However, the beneficial effects provided by probiotics depend on regular and recommended daily intake.

It is recommended to have a probiotic cell count in the range of $10^8$ - $10^9$ CFU.g$^{-1}$ in the feed. Based on this count, it is possible to ensure that, after ingestion, at least $10^6$ - $10^7$ CFU.g$^{-1}$ can reach the specific site of action (small and/or large intestine) and, thus, promote beneficial health effects (Ahmad et al., 2018). One of the major problems faced by the food industry is caused by the low survival of probiotics when exposed to different factors, such as processing steps, long storage periods, and adverse conditions during passage through the gastrointestinal system (Shori 2017).

Thus, among several techniques aimed at maintaining the cellular viability of these microorganisms, microencapsulation represents one of the most efficient strategies. The microencapsulation process protects probiotic cells against adverse conditions, thus ensuring their presence in sufficient quantities to exert their beneficial effects (Calabuig-Jiménez et al., 2019). Various techniques can be employed for the microencapsulation of probiotics, with complex coacervation standing out.

The complex coacervation process involves the presence of two or more polymers with opposite charges that form coacervates through interaction or complexation driven mainly by electrostatic interaction, under appropriate conditions (Schmitt and Turgeon 2011). Among the existing studies related to the complex coacervation process, the oldest was reported by Bungenberg et al., (1949) a century ago. Currently, the complex coacervation technique using biopolymers is widely used in several applications. Recently, one of the applications that have gained prominence in the food, cosmetic and pharmaceutical industries involve the protection and release of compounds under adverse environmental conditions (Muhoza et al., 2022).

For the microencapsulation of probiotics, the use of the complex coacervation process offers advantages as it uses mild conditions and does not require the use of organic solvents or high temperatures, thus representing an interesting alternative (Da Silva et al., 2019). Thus, the proposed chapter focuses on presenting the main characteristics of the complex coacervation process. Within the proposed theme, there will be an approach to the challenges faced by the food industry and the use of microencapsulation by complex coacervation as an alternative for maintaining the stability and survival of these microorganisms during processing, storage, and passage through the gastrointestinal tract.

## Coacervation

Historically, the word "coacervate" has its origin in the Latin word "acervus," meaning a pile or aggregate. The prefix "co" translates to the union of colloidal particles (Menger and Sykes 1998). Bringing it to the present day, the term "coacervation" represents the process of phase separation by modifying the medium (such as pH, ionic strength, temperature, and solubility) under controlled conditions. During this procedure, phase separation occurs in a colloidal dispersion. These phases are immiscible: one contains a high concentration of colloid (the coacervate phase), and the other is dilute, containing small amounts of colloid (the equilibrium phase).

According to the classification of the process, there is simple and complex coacervation. Simple coacervation involves using one type of macromolecule and can be induced by adding alcohols, salts, or increasing the temperature. In this process, phase separation results in a phase with a high macromolecule concentration in equilibrium with the second phase containing a higher concentration of water (Eghbal and Choudhary 2018). Complex coacervation occurs when two or more types of macromolecules with opposite charges combine through multiple interactions, primarily electrostatic. As a result of the interactions between the macromolecules, coacervate complexes form, causing a separation of phases in which one phase is rich in solvent and the other contains both polymers used (Eghbal et al., 2016). Complex coacervates tend to have a denser structure and exhibit greater thermal stability and resistance to mechanical shear (Wan et al., 2023b).

Tiebackx (1911) was the first researcher to report the phenomenon of phase separation involving the use of macromolecules with opposite charges. However, Bungenberg de Jong and Kruyt (1929) were the first to investigate the complex coacervation process using gum Arabic and gelatin. Gelatin, a water-soluble protein, and gum arabic, a long-chain polymer, both with high molecular weight, constitute the first pair used and probably the most studied (Moulik et al., 2022). The combination of gelatin and gum arabic provides efficient surface properties and cold solubility due to the presence of charged residual groups and peptide fragments (Phillips, Takigami, and Takigami 1996).

According to the results obtained by Bungenberg de Jong and Kruyt (1929), the researchers verified that the interaction between the biopolymers was electrostatic, and the coacervation was strongly dependent on the molecular charge density, pH, and ionic strength. Currently, among the macromolecules used in the complex coacervation process, gelatin still represents the most widely used and marketed protein for this purpose. The use by the industry began in the 50s, and since then, many works reporting the microencapsulation of different compounds using complex coacervation can be found. The complex coacervation process in the microencapsulation of food ingredients mainly uses proteins and polysaccharides, where the complex coacervates envelop the core material, forming the microcapsules (Timilsena et al., 2019).

## Complex Coacervation

As mentioned earlier, the complex coacervation process is based on the electrostatic interaction between oppositely charged molecules at a specific pH. Consequently, complex coacervates are formed, leading to phase separation. One phase becomes enriched with the solvent, while the other contains both polymers employed (Yang et al., 2015). In the food industry, proteins and anionic polysaccharides are the primary polymers utilized in this process, although various combinations have been described in the literature (Gomez-Estaca et al., 2016). Some notable options include gelatin, gum arabic, alginate, albumin, casein, agar, gums, pectins, and chitosan (Rutz et al., 2017).

During the formation of complex coacervates, the electrostatic interaction between biopolymers with opposite charges results in the formation of soluble complexes. These complexes then aggregate to reduce the free energy of the system until they reach a size and surface properties that render them insoluble (Eratte et al., 2018). However, certain conditions in the surrounding medium, such as high salt content, can suppress electrostatic interactions. In such cases, other interactions such as hydrogen bonds and hydrophobic interactions may contribute to the formation of complex coacervates (Turgeon, Schmitt, and Sanchez 2007). Various factors influence the process of complex coacervation. For instance, the properties of the polymers used, such as the type and position of charged groups, as well as the overall charge, play a significant role (Comunian et al., 2022). Additionally, environmental conditions including the type and strength of salt ions, pH, protein/polysaccharide ratio, total polymer concentration, and temperature, also impact the complex coacervation process. Figure 1 illustrates phase separation during the complex coacervation process.

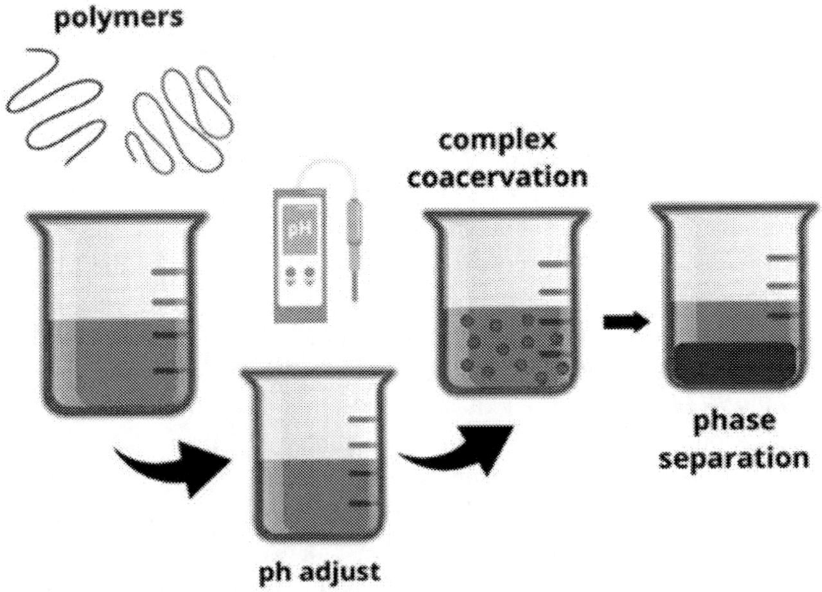


**Figure 1**. Complex coacervation and system phase separation.

## Complex Coacervate Characterization Methods

To increase understanding of the formation and utilization of protein/polysaccharide complex coacervates, several studies and research have been conducted to evaluate the characteristics of these structures. Consequently, in recent years, the function, structure, and conformation of protein/polysaccharide complex coacervate have been the focus of interest for various researchers (Wan et al., 2023a).

In addition, three approaches are mainly used for the optimization of the complex coacervation process: surface charge density (zeta potential), turbidity as a function of pH and the mass ratio of the polymers used, and yield (Huang et al., 2012; Timilsena et al., 2015). The mass ratio of the molecules used, and the pH are critical factors that affect the attraction and consequently the complexation (Xiao, Gu, and Tan 2014). To determine the pH value at which the maximum interaction between biopolymers occurs, that is, the neutralization of charges, it is necessary to measure the electrophoretic mobility of the molecules, from which the zeta potential can be calculated.

The use of UV-Vis spectrophotometer has as its main objective the measurement of turbidity in polymer mixtures, as well as the formation of complex coacervates. This technique allows for the assessment of turbidity increase due to the production of complex coacervates (Eghbal and Choudhary 2018). However, depending on the conditions present in the medium, it is possible to reach a plateau in the turbidity value, which may indicate the presence of electrostatic interactions between all possible negative and positive charges of the biopolymers. Consequently, this results in the coalescence of all complex coacervates and the formation of the insoluble coacervate phase.

Paula et al., (2019) used zeta potential analysis and UV-visible spectrophotometry to optimize the conditions for the microencapsulation of *Lactobacillus plantarum* probiotic cells through a dual process consisting of emulsification followed by complex coacervation with gelatin and gum arabic. Liu and Jiang (2023) used the dry weight of the obtained coacervates to evaluate the yield and determine the maximum degree of coacervation. Thus, the authors found that the lowest yield obtained was 59.3% when the gelatin/pectin mixing ratio was 1:1. However, an increase in yield was observed with an increase in the proportion of gelatin, reaching a maximum of 71.2% at the optimal mixing ratio (G:P = 3:1). At low concentrations of gelatin/pectin, the yield of coacervates decreased significantly due to the formation of soluble complexes instead of complex coacervates and the lower charge neutralization between the two biopolymers (Timilsena et al., 2016).

Comunian et al., (2022) evaluated the coacervation process using two types of Arabic gum (AG) (Acacia Senegal-GA1 and Acacia seyal-GA2) with soluble pea protein (SPP). They employed techniques such as Zeta potential, turbidity, morphology, SPP secondary structure, UV/Vis absorbance, and thermodynamic parameters. Based on the obtained results, it was possible to observe that the highest coacervation yield occurred at SPP: AG ratio of 3:1 (w/w) and pH 3.5-4.0, with alterations in the SPP secondary structure. The negative values of Gibbs free energy indicated a spontaneous process. The use of Acacia Senegal-GA1 resulted in stronger interactions with SPP, offering new alternatives for the encapsulation of bioactive compounds.

Carpentier et al., (2021) analyzed the possible formation of coacervate complexes between pea protein isolates (PPI) and gum using turbidity measurements and zeta potential. Regarding the pH of soluble complex formation, the pH of the complex coacervates increased with the increase in the ratio between protein and anionic polysaccharide. The results showed that tragacanth gum can be used as an alternative to Arabic gum for forming complex coacervates with PPI based on zeta potential measurements and coacervation yield studies.

## Microencapsulation by Complex Coacervation Process

During the microencapsulation technique, a process known as "packaging" takes place, where an active ingredient is coated with a wall material. As a result, it provides protection for unstable molecules/compounds against interactions and adverse conditions during processing, storage, and passage through the gastrointestinal tract.

The complex coacervation technique represents a promising technology due to its high capacity for encapsulating compounds and the possibilities of controlled release based on mechanical tension, temperature, or sustained release. During the microencapsulation process using complex coacervation, phase separation of the system occurs, followed by the deposition of complex coacervates around the suspended active ingredient in the medium (Gouin 2004). The microcapsules obtained through the complex coacervation process have core-shell structures, either mononuclear, polynuclear, or aggregated like "grape clusters." In Figure 2 it is possible to verify the complex coacervation process being used for the microencapsulation of probiotic cells.

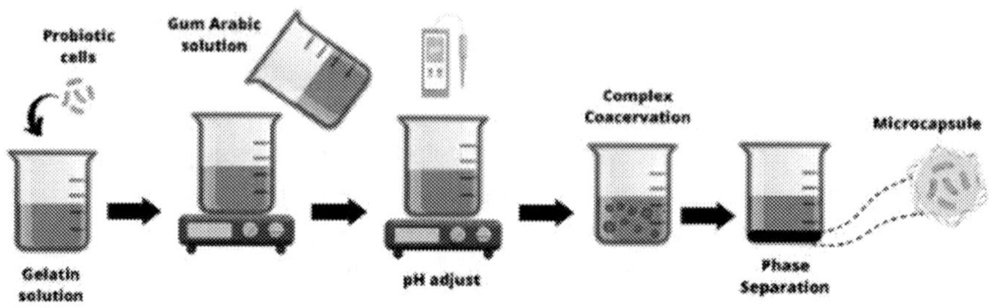


**Figure 2.** Probiotics encapsulation by complex coacervation.

The instability of microcapsules, under certain conditions, due to the ionic nature of the electrostatic interaction between polymers represents a disadvantage of the process. Consequently, several modification techniques are employed aiming to improve the intermolecular interaction. Among the processes used, chemical and enzymatic crosslinking are highlighted. The majority of chemical crosslinking agents mentioned in the literature, such as formaldehyde, glutaraldehyde, glyoxal, diisocyanate, and epichlorohydrin, have limited applications due to their toxicity and the difficulty of complete removal with a suitable solvent to achieve biological acceptance (Butstraen and Salaün 2014). As a result, other compounds such as Ca2+, tannic acid, oxidized tannic acid, sodium tripolyphosphate, transglutaminase enzyme, and citric acid have been reported to help stabilize and modify microstructures of coacervate complexes, functioning as crosslinking agents (Guo et al., 2022).

The introduction of crosslinks via enzymatic means can be an alternative. For example, the use of transglutaminase enzyme acts by forming intra and intermolecular crosslinks between two residual amino acids present in the protein structure. These crosslinks improve rheological and physical properties, bringing significant changes to the protein molecule without negatively affecting the sensory and nutritional qualities of the product (Taghi Gharibzahedi et al., 2018). Currently, there are also investigations involving the use of transglutaminase to improve the characteristics of microcapsules containing probiotics.

Marques da Silva et al., (2021) evaluated the effect of transglutaminase crosslinking on the resistance of microcapsules containing Lactobacillus acidophilus produced by complex coacervation. Additionally, they evaluated the probiotic viability under different adverse conditions, such as high temperatures and simulated gastrointestinal conditions. The authors found that the use of transglutaminase maintained the structure of the rounded, multinucleated microcapsules and showed a homogeneous distribution of probiotic cells. Regarding thermal resistance, the crosslinked microcapsules exhibited greater protection, with viable cell counts of up to 10 log CFU g−1, approximately. Under simulated gastrointestinal conditions, encapsulation combined with crosslinking showed promising results, with the dry form being more efficient in protection, particularly the treatment with a higher amount of transglutaminase (9.07 log CFU g−1).

Peñalva et al., (2023) evaluated the application of the complex coacervation process in the microencapsulation of probiotic cells. To do so, the authors performed the ionic interaction between sodium caseinate and chitosan. The obtained microcapsules were crosslinked using different agents, including vanillin, tripolyphosphate TPP, or calcium salts, and dried by spray drying. Dong et al., (2008) developed spherical multinucleated microcapsules of gum Arabic containing peppermint oil through the complex coacervation process. Transglutaminase (TG) was used to harden the microcapsules, replacing the use of traditional reagents such as formaldehyde or glutaraldehyde. The effect of various crosslinking parameters on the efficacy of transglutaminase (TG) utilization was investigated. According to the obtained results, the best parameters were hardening for 6 hours at 15°C and pH 6.0 with a TG concentration of 15 U/g gelatin. Compared to the use of formaldehyde, transglutaminase showed similar effectiveness.

Chen et al., (2012) characterized the properties of complex coacervates obtained through electrostatic interaction between WPI and beet pectin and crosslinking with laccase. For this purpose, the interaction between whey protein isolates and beet pectin in aqueous solutions was studied at different pH levels (3–7). Zeta potential and light scattering techniques were used to provide information about the electrical charge and biopolymer aggregation. The obtained results showed that crosslinking had an influence on the physical properties of complex coacervates and formed structures that could provide convenient means for the food industry to incorporate bioactive components into food products.

## Probiotics Encapsulation by Complex Coacervation

Probiotics are live microorganisms that, when administered in adequate amounts, can result in health benefits for those who consume them (FAO/WHO). However, to obtain these benefits, it is important to maintain cell viability and ensure that microorganisms reach their specific point of action in appropriate quantities. Therefore, the incorporation of probiotics into food matrices has been widely studied and evaluated. As a result, the functional food industry has grown rapidly. In recent research, the global functional food market was valued at over $260 billion and is projected to exceed $380 billion in sales by 2024 (Transparency Market Research, 2018/2026).

However, maintaining the cell viability of certain probiotic strains still poses challenges, considering the physicochemical characteristics of the food matrix, such as low pH, presence of oxygen, and storage temperature. Research and studies have shown that microencapsulation can provide physical protection and maintenance of probiotic cell viability in food. Additionally, the process has been utilized by the food industry as a means to add value to their products and differentiate themselves in the market (Da Silva et al., 2019).

The microcapsules produced by complex coacervation and containing the probiotic cells are generally dried. This process occurs to facilitate protection and storage. Furthermore, probiotic cells are less stable in liquid form than in dry form, therefore different methods of dehydration, e.g., spray drying, spouted bed, freeze-drying, etc. For the microencapsulation of probiotics, complex coacervation has proven to be an extremely relevant technique due to its high encapsulation efficiency and protection. Furthermore, microcapsules produced through coacervation are versatile and exhibit excellent release characteristics modulated by changes in ionic strength, pH, and temperature (Marques et al., 2018). Thus, the release of encapsulated microorganisms can be controlled by pH (Rutz et al., 2017), providing a wide opportunity and perspective for the development of probiotic foods.

According to Peñalva et al., (2023), the level of protection is an important aspect that should be evaluated in probiotic cell delivery systems. Satisfactory protection results ensure the survival of microorganisms in the intestinal environment for a sufficient period, allowing the cells to reach intestinal surfaces where they colonize and provide health benefits. The wall materials used during the microencapsulation process play a crucial role in determining the size, morphology, texture, porosity, and other relevant properties of the capsules. Furthermore, depending on the materials used, it is possible to influence the protection or targeted delivery of probiotics. The selection of wall materials and the adopted technologies in the manufacturing of probiotic microcapsules are of paramount importance as they strictly reflect the morphological and functional properties of the capsules (Eratte et al., 2018).

Matos-Jr et al., (2019) studied the resistance, viability, and functionality of two strains of probiotics immobilized in solid lipid microparticles coated with a complex of gelatin and gum arabic. The microcapsules were evaluated for size, morphology, and resistance under stress conditions. According to the results, the authors found that encapsulation improved the stability of these microorganisms in the presence of salt and gastrointestinal conditions. Paula et al., (2019), when microencapsulating probiotic cells of *Lactobacillus plantarum* through a dual process consisting of emulsification followed by complex coacervation with gelatin and gum arabic, found that the viability of the encapsulated *L. plantarum* cells was 8.6 log $CFU \cdot g^{-1}$ and the encapsulation efficiency was 97.78%. After simulating in vitro gastrointestinal conditions, the viability of the encapsulated cells was 80.4%, while for the free cells, it was only 25.0%.

Peñalva et al., (2023) studied and evaluated the potential of casein-chitosan microparticles, prepared by complex coacervation, to encapsulate *L. plantarum*. According to the obtained results, there was an increase in the viability of the microencapsulated bacteria during processing and storage. Additionally, the ability to reach the distal areas of the intestine and stimulate the immune system in vivo also showed better results. Ribeiro et al., (2014) evaluated the application of incorporating microencapsulated *Lactobacillus acidophilus* in the development of stirred yogurt. For this purpose, *L. acidophilus* cells were microencapsulated using ion gelation and complex coacervation techniques using pectin and whey protein. According to the results obtained, it was found that the microencapsulation process of *L. acidophilus* provided lower post-acidification in probiotic yogurts and protection to the microorganism during 35 days of refrigerated storage and passage through simulated gastrointestinal conditions.

Marques da Silva et al., (2021) developed microcapsules containing *Lactobacillus acidophilus* through complex coacervation followed by crosslinking with transglutaminase and evaluated the effect of their addition in different fruit juices. The authors found that the coacervation process associated with enzymatic crosslinking protected *L. acidophilus* in different fruit juices, demonstrating a suitable methodology for adding probiotics to this challenging food matrix. It ensured the survival of *L. acidophilus* for up to 63 days, generating products with innovative and promising probiotic appeal.

Sharifi et al., (2021) used whey protein isolate (WPI) and gum arabic (GA) as wall materials to encapsulate *Lactobacillus plantarum* ATCC 8014 (P) alone or in combination with phytosterols (Ph) using the complex coacervation technique. The obtained data showed that the cell viability of *L.*

*plantarum* ATCC 8014 in both liquid and dry microcapsules (freeze-dried and spray-dried) was enhanced in the presence of phytosterols compared to their absence, and the survival capacity of *L. plantarum* ATCC 8014 in freeze-dried microcapsules was superior to spray-dried microcapsules during the storage time (60 days).

## Conclusion and Prospects

During processing, food storage, and passage through the gastrointestinal tract, probiotic cells can lose their viability. Therefore, complex coacervation represents an alternative to maintain or improve the viability of probiotic cells in the face of these adverse conditions. Studies and research have reported that complex coacervation is a simple, cheaper, reproducible, and more efficient method for microencapsulation. Additionally, microencapsulation using complex coacervation can be performed under mild processing conditions without the need for toxic solvents, offering advantages for the microencapsulation of probiotic cells.

However, for application in the field of probiotic microencapsulation, association with a drying method is necessary. For example, freeze-drying or spray drying, to achieve high cell counts of probiotic cells and facilitate handling, storage, and transportation. In this sense, the food industry must analyze and consider various aspects, such as suitable coating materials for controlled release, the conditions of the medium to be used, production feasibility to meet consumer specifications and manufacturing costs.

In general, complex coacervation is an emerging microencapsulation technique for probiotic cells considered advantageous. This technology can enhance the addition of probiotic cells in different foods and thus achieve the goals of this growing market. Studies on the process and specific research focusing on the structure of microcapsules obtained through complex coacervation are vital for the design of new probiotic foods. Regarding the analysis of the incorporation of these microcapsules, there is a need for research aiming to evaluate possible interactions with the food matrix. Consequently, a better understanding of this process can optimize the process, reduce costs, and achieve the necessary scalability for industrial purposes.

## References

Ahmad, Bilal, Asir Gani, Adil Gani, Asima Shah, and Farooq Ahmad. 2018. "Production of RS4 from Rice Starch and Its Utilization as an Encapsulating Agent for Targeted Delivery of Probiotics." *Food Chemistry* 239: 287–94.https://doi.org/10.1016/j.foodchem.2017.06.110

Bungenberg de Jong, H. G., Kruyt, H. R. "Colloid Science," Elsevier, Amsterdam, 1949, pp. 232-258.

Butstraen, Chloé, and Fabien Salaün. 2014. "Preparation of Microcapsules by Complex Coacervation of Gum Arabic and Chitosan." *Carbohydrate Polymers* 99: 608–16. https://doi.org/10.1016/j.carbpol.2013.09.006

Calabuig-jiménez, Laura, Ester Betoret, Noelia Betoret, Francesca Patrignani, Cristina Barrera, Lucía Seguí, Rosalba Lanciotti, and Marco Dalla. 2019. "High Pressures Homogenization (HPH) to Microencapsulate L . Salivarius Spp. Salivarius in Mandarin Juice. Probiotic Survival and in Vitro Digestion." *Journal of Food Engineering* 240 (July 2018): 43–48. https://doi.org/10.1016/j.jfoodeng.2018.07.012

Carpentier, Jérémy, Egle Conforto, Carine Chaigneau, Jean Eudes Vendeville, and Thierry Maugard. 2021. "Complex Coacervation of Pea Protein Isolate and Tragacanth Gum: Comparative Study with Commercial Polysaccharides." *Innovative Food Science and Emerging Technologies* 69 (September 2020). https://doi.org/10.1016/j.ifset.2021.102641

Chen, Bingcan, Hongjun Li, Yangping Ding, and Huayi Suo. 2012. "Formation and Microstructural Characterization of Whey Protein Isolate/beet Pectin Coacervations by Laccase Catalyzed Cross-Linking." *Lwt* 47 (1): 31–38. https://doi.org/10.1016/j.lwt.2012.01.006

Comunian, Talita A., Artwin Archut, Laura G. Gomez-Mascaraque, André Brodkorb, and Stephan Drusch. 2022. "The Type of Gum Arabic Affects Interactions with Soluble Pea Protein in Complex Coacervation." *Carbohydrate Polymers* 295 (February). https://doi.org/10.1016/j.carbpol.2022.119851

Dong, Zhi-jian, Shu-qin Xia, Serenus Hua, and Khizar Hayat. 2008. "Optimization of Cross-Linking Parameters during Production of Transglutaminase-Hardened Spherical Multinuclear Microcapsules by Complex Coacervation" 63: 41–47. https://doi.org/10.1016/j.colsurfb.2007.11.007

Eghbal, Noushin, and Ruplal Choudhary. 2018. "Complex Coacervation: Encapsulation and Controlled Release of Active Agents in Food Systems." *Lwt* 90 (December 2017): 254–64. https://doi.org/10.1016/j.lwt.2017.12.036

Eghbal, Noushin, Mohammad Saeid Yarmand, Mohammad Mousavi, Pascal Degraeve, Nadia Oulahal, and Adem Gharsallaoui. 2016. "Complex Coacervation for the Development of Composite Edible Films Based on LM Pectin and Sodium Caseinate." *Carbohydrate Polymers* 151: 947–56. https://doi.org/10.1016/j.carbpol.2016.06.052

Eratte, Divya, Kim Dowling, Colin J. Barrow, and Benu Adhikari. 2018. "Recent Advances in the Microencapsulation of Omega-3 Oil and Probiotic Bacteria through Complex Coacervation: A Review." *Trends in Food Science and Technology* 71 (September 2016): 121–31. https://doi.org/10.1016/j.tifs.2017.10.014

Gomez-Estaca, J., T. A. Comunian, P. Montero, R. Ferro-Furtado, and C. S. Favaro-Trindade. 2016. "Encapsulation of an Astaxanthin-Containing Lipid Extract from Shrimp Waste by Complex Coacervation Using a Novel Gelatin-Cashew Gum Complex." *Food Hydrocolloids* 61: 155–62. https://doi.org/10.1016/j.foodhyd.2016.05.005

Gouin, Sébastien. 2004. "Microencapsulation: Industrial Appraisal of Existing Technologies and Trends." *Trends in Food Science and Technology* 15 (7–8): 330–47. https://doi.org/10.1016/j.tifs.2003.10.005

Guo, Qi, Shuai Li, Gengan Du, Hong Chen, Xiaohai Yan, Shuaidan Chang, Tianli Yue, and Yahong Yuan. 2022. "Formulation and Characterization of Microcapsules Encapsulating Carvacrol Using Complex Coacervation Crosslinked with Tannic Acid." *Lwt* 165 (June): 113683. https://doi.org/10.1016/j.lwt.2022.113683

Huang, Guo Qing, Yan Ting Sun, Jun Xia Xiao, and Jian Yang. 2012. "Complex Coacervation of Soybean Protein Isolate and Chitosan." *Food Chemistry* 135 (2): 534–39. https://doi.org/10.1016/j.foodchem.2012.04.140.

Lilly, Daniel M., and Rosalie H. Stillwell. 1964. "Probiotics: Growth-Promoting Factors Produced by Microorganisms." *Science* 147 (3659): 747–48.

Liu, Yunshan, and Jianzhong Jiang. 2023. "Preparation of β-Ionone Microcapsules by Gelatin/pectin Complex Coacervation." *Carbohydrate Polymers* 312 (November 2022): 120839. https://doi.org/10.1016/j.carbpol.2023.120839

Marques, Thaiane, Eduardo Jacob, Cristiane Franco, Alexandre José, Érico Marlon, De Moraes Flores, Mariana Heldt, Cristiane de Bona da Silva, Carlos Raimundo Ferreira Grosso, Cristiano Ragagnin de Menezes. 2018. "Development and Characterization of Microcapsules Containing *Bifidobacterium* Bb-12 Produced by Complex Coacervation Followed by Freeze Drying." *LWT - Food Science and Technology* 90 (August 2017): 412–17. https://doi.org/10.1016/j.lwt.2017.12.057

Marques da Silva, Thaiane, Vandré Sonza Pinto, Vítor Ramires Fonseca Soares, Débora Marotz, Alexandre José Cichoski, Leila Queiroz Zepka, Eduardo Jacob Lopes, Cristiane de Bona da Silva, and Cristiano Ragagnin de Menezes. 2021a. "Viability of Microencapsulated *Lactobacillus Acidophilus* by Complex Coacervation Associated with Enzymatic Crosslinking under Application in Different Fruit Juices." *Food Research International* 141 (October 2020). https://doi.org/10.1016/j.foodres.2021.110190

Matos-Jr, Fernando Eustáquio de, Marluci Palazzolli da Silva, Marcia Gabriela Consiglio Kasemodel, Tizá Teles Santos, Patricia Burns, Jorge Reinheimer, Gabriel Vinderola, and Carmen Sílvia Favaro-Trindade. 2019. "Evaluation of the Viability and the Preservation of the Functionality of Microencapsulated *Lactobacillus Paracasei* BGP1 and *Lactobacillus Rhamnosus* 64 in Lipid Particles Coated by Polymer Electrostatic Interaction." *Journal of Functional Foods* 54 (December 2018): 98–108. https://doi.org/10.1016/j.jff.2019.01.006

Menger, Fredric M., and Bridget M. Sykes. 1998. "Anatomy of a Coacervate." *Langmuir* 14 (15): 4131–37. https://doi.org/10.1021/la980208m.

Moulik, Satya Priya, Animesh Kumar Rakshit, Animesh Pan, and Bappaditya Naskar. 2022. "An Overview of Coacervates: The Special Disperse State of Amphiphilic and

Polymeric Materials in Solution." *Colloids and Interfaces* 6 (3). https://doi.org/10.3390/colloids6030045

Muhoza, Bertrand, Shuqin Xia, Xuejiao Wang, Xiaoming Zhang, Yang Li, and Shuang Zhang. 2022. "Microencapsulation of Essential Oils by Complex Coacervation Method: Preparation, Thermal Stability, Release Properties, and Applications." *Critical Reviews in Food Science and Nutrition* 62 (5): 1363–82. https://doi.org/10.1080/10408398.2020.1843132

Paula, Daniele de Almeida, Eliane Maurício Furtado Martins, Nataly de Almeida Costa, Patrícia Martins de Oliveira, Eduardo Basílio de Oliveira, and Afonso Mota Ramos. 2019a. "Use of Gelatin and Gum Arabic for Microencapsulation of Probiotic Cells from *Lactobacillus Plantarum* by a Dual Process Combining Double Emulsification Followed by Complex Coacervation." *International Journal of Biological Macromolecules* 133: 722–31. https://doi.org/10.1016/j.ijbiomac.2019.04.110

Peñalva, Rebeca, Ana Luisa Martínez-López, Carlos Gamazo, Carlos J. Gonzalez-Navarro, Carolina González-Ferrero, Raquel Virto-Resano, Ana Brotons-Canto, Ana Isabel Vitas, Maria Collantes, Ivan Peñuelas, Juan M. Irache. 2023. "Encapsulation of *Lactobacillus Plantarum* in Casein-Chitosan Microparticles Facilitates the Arrival to the Colon and Develops an Immunomodulatory Effect." *Food Hydrocolloids* 136 (July 2022). https://doi.org/10.1016/j.foodhyd.2022.108213

Phillips, G.O., S. Takigami, and M. Takigami. 1996. "Hydration Characteristics of the Gum Exudate from Acacia Senegal." *Food Hydrocolloids* 10 (1): 11–19. https://doi.org/10.1016/S0268-005X(96)80048-8

Ribeiro, Maria Cecília E., Karina S. Chaves, Clarice Gebara, Flávia N.S. Infante, Carlos R.F. Grosso, and Mirna L. Gigante. 2014. "Effect of Microencapsulation of *Lactobacillus Acidophilus* LA-5 on Physicochemical, Sensory and Microbiological Characteristics of Stirred Probiotic Yoghurt." *Food Research International* 66: 424–31. https://doi.org/10.1016/j.foodres.2014.10.019

Rutz, Josiane K., Caroline D. Borges, Rui C. Zambiazi, Michele M. Crizel-Cardozo, Luiza S. Kuck, and Caciano P.Z. Noreña. 2017. "Microencapsulation of Palm Oil by Complex Coacervation for Application in Food Systems." *Food Chemistry* 220: 59–66. https://doi.org/10.1016/j.foodchem.2016.09.194

Schmitt, Christophe, and Sylvie L Turgeon. 2011. "Protein / Polysaccharide Complexes and Coacervates in Food Systems." *Advances in Colloid and Interface Science* 167 (1–2): 63–70. https://doi.org/10.1016/j.cis.2010.10.001

Sharifi, Sohrab, Mahmoud Rezazad-Bari, Mohammad Alizadeh, Hadi Almasi, and Saber Amiri. 2021. "Use of Whey Protein Isolate and Gum Arabic for the Co-Encapsulation of Probiotic *Lactobacillus Plantarum* and Phytosterols by Complex Coacervation: Enhanced Viability of Probiotic in Iranian White Cheese." *Food Hydrocolloids* 113 (December 2020): 106496. https://doi.org/10.1016/j.foodhyd.2020.106496

Shori, Amal Bakr. 2017. "Microencapsulation Improved Probiotics Survival During Gastric Transit." *HAYATI Journal of Biosciences* 24 (1): 1–5. https://doi.org/10.1016/j.hjb.2016.12.008

Silva, Thaiane Marques Da, Juliano Smanioto Barin, Eduardo Jacob Lopes, Alexandre José Cichoski, Erico Marlon De Moraes Flores, Cristiane De Bona Da Silva, and Cristiano Ragagnin De Menezes. 2019. "Development, Characterization and Viability Study of

Probiotic Microcapsules Produced by Complex Coacervation Followed by Freeze-Drying." *Ciencia Rural* 49 (7). https://doi.org/10.1590/0103-8478cr20180775

Song, Xiaoran, Yanan Liu, Xin Zhang, Peifang Weng, Ruilin Zhang, and Zufang Wu. 2023. "Role of Intestinal Probiotics in the Modulation of Lipid Metabolism: Implications for Therapeutic Treatments." *Food Science and Human Wellness* 12 (5): 1439–49. https://doi.org/10.1016/j.fshw.2023.02.005

Taghi Gharibzahedi, Seyed Mohammad, Mohamed Koubaa, Francisco J. Barba, Ralf Greiner, Saji George, and Shahin Roohinejad. 2018. "Recent Advances in the Application of Microbial Transglutaminase Crosslinking in Cheese and Ice Cream Products: A Review." *International Journal of Biological Macromolecules* 107: 2364–74. https://doi.org/10.1016/j.ijbiomac.2017.10.115

Timilsena, Yakindra Prasad, Raju Adhikari, Colin J. Barrow, and Benu Adhikari. 2016. "Microencapsulation of Chia Seed Oil Using Chia Seed Protein Isolate-Chia Seed Gum Complex Coacervates." *International Journal of Biological Macromolecules* 91: 347–57. https://doi.org/10.1016/j.ijbiomac.2016.05.058

Timilsena, Yakindra Prasad, Taiwo O Akanbi, Nauman Khalid, Benu Adhikari, and Colin J Barrow. 2019. "Complex Coacervation : Principles, Mechanisms, and Applications in Microencapsulation." *International Journal of Biological Macromolecules* 121: 1276–86. https://doi.org/10.1016/j.ijbiomac.2018.10.144

Timilsena, Yakindra Prasad, Bo Wang, Raju Adhikari, and Benu Adhikari. 2015. "Preparation and Characterization of Chia Seed Protein Isolate-Chia Seed Gum Complex Coacervates." *Food Hydrocolloids* 52: 554–63. https://doi.org/10.1016/j.foodhyd.2015.07.033

Transparency Market Research, Transparency Market Research 2018-2026. https://www.transparencymarketresearch.com/probiotics-market.html

Turgeon, S. L., C. Schmitt, and C. Sanchez. 2007. "Protein-Polysaccharide Complexes and Coacervates." *Current Opinion in Colloid and Interface Science* 12 (4–5): 166–78. https://doi.org/10.1016/j.cocis.2007.07.007

Wan, Xiaoshan, Meihui Zhao, Mengxue Guo, Peng Li, Haohao Shi, Xueying Zhang, Zhongyuan Liu, and Guanghua Xia. 2023a. "Characterization of Coacervation Behavior between Whey Protein Isolate and Gum Arabic: Effects of Heat Treatment." *Food Chemistry: X* 18 (May): 100703. https://doi.org/10.1016/j.fochx.2023.100703

Xiaoshan Wan, Meihui Zhao, Mengxue Guo, Peng Li, Haohao Shi, Xueying Zhang, Zhongyuan Liu, Guanghua Xia 2023. "Food Chemistry : X Characterization of Coacervation Behavior between Whey Protein Isolate and Gum Arabic : Effects of Heat Treatment." *Food Chemistry: X* 18 (April): 100703. https://doi.org/10.1016/j.fochx.2023.100703

Xiao, Qian, Xiaohong Gu, and Suo Tan. 2014. "Drying Process of Sodium Alginate Films Studied by Two-Dimensional Correlation ATR-FTIR Spectroscopy." *Food Chemistry* 164: 179–84. https://doi.org/10.1016/j.foodchem.2014.05.044

Yang, Xun, Na Gao, Liandong Hu, Jianli Li, and Yongbing Sun. 2015. "Development and Evaluation of Novel Microcapsules Containing Poppy-Seed Oil Using Complex Coacervation." *Journal of Food Engineering* 161: 87–93. https://doi.org/10.1016/j.jfoodeng.2015.03.027

## Chapter 8

# Coated Particles for Improving Probiotic Protection

**Karen Cristina Guedes Silva[1,*], PhD**
**and Flávia Souza Almeida[2], PhD**
[1]Federal Institute of Paraná – IFPR, Assis Chateaubriand, Paraná, Brazil
[2]Department of Food Engineering and Technology, School of Food Engineering, University of Campinas – UNICAMP, Campinas, São Paulo, Brazil

## Abstract

Coating layers can largely impact the protection of probiotics, as they can help to overcome the weakness of single-layer particles such as the high porosity and permeability, and poor mucoadhesiveness, besides the great sensitivity to chelating agents, which promotes disintegration and release of probiotics cells at acidic media. Depending on the coating and/or core material composition, good protection of the probiotics can be achieved, improving the defense of cells against adverse conditions of the digestive tract, to oxygen and storage temperatures. Moreover, additional layers may also promote target release and great mucoadhesion of probiotics to the intestinal epithelial cells. Coating layers are based on natural compounds such as proteins, polysaccharides, and lipids, and the most common methods used in the manufacture of multilayer particles include co-extrusion, double emulsion, layer-by-layer, fluidized bed coater, and electrospray coating. Although these techniques are very promising, they still present a lack of control over the thickness of the coating layer. Despite some reports in the literature have highlighted the positive impact of adding multiple layers in the manufacture of microparticles, it is important to stress that the desired number of coating layers is related

* Corresponding Author's Email: karen.silva@ifpr.edu.br.

In: Microencapsulation of Probiotics
Editors: D. de Almeida Paula, K. Cristina Guedes Silva and N. de Almeida Costa
ISBN: 979-8-88697-979-4

to different parameters such as the microorganism, the conditions the probiotics are submitted to and the time-frame of evaluation. As further discussed, increasing the number of coating layers does not always increase probiotic viability. This chapter, therefore, aimed to present an overview of multilayer microparticles, considering the different materials and techniques applied, emphasizing the morphology and release properties of the multilayer microencapsulated probiotics.

**Keywords:** coating layer, probiotic, biopolymer, multilayer particle

## Introduction

Microencapsulation has been a successful technique for enhancing the survival of probiotic microorganisms against critical conditions. The protection efficiency provided by microencapsulation is dependent on different parameters, especially the microorganism strain, the method used, and the coating material (Pech-Canul et al. 2020). Moreover, small changes in the composition of coating and/or core material greatly impact the final characteristics of the microparticles (Menezes et al. 2019).

Innumerous biopolymers can be used for microencapsulation of probiotics and they are mainly based on proteins, polysaccharides, and lipids. Examples comprise alginate (Cezarino et al. 2022), chitosan (Phuong Ta et al. 2021), starch (Dafe et al. 2017), pectin (Menezes et al. 2019), whey protein (Mabrouk et al. 2021) and gelatin (Albadran et al. 2020), besides some lipids such as cocoa butter (Pedroso et al. 2013), and hydrogenated vegetable oils (Arslan-Tontul and Erbas 2017). Different mechanisms can be involved in the protection of the probiotics by using these materials, such as through the control of stress response against gastric environmental conditions and other external variables (e.g., UV light and heat) (Pech-Canul et al. 2020).

Because a great number of microparticles only present a single layer of protection around probiotics, it sometimes results in poor protection when compared to particles fabricated by the multilayer approach. Coating layers are known for improving mucoadhesion, the target release of nutrients, and the promotion of limited permeability, which in turn leads toward a reduction to oxygen exposure by the encapsulated bioactive such as probiotics, as well as improved viability at the acidic environment for the latter ones (Talebian et al. 2022). In particular, uncoated alginate particles can be very porous and sensitive to chelating agents (e.g., phosphate or citrate), being responsible for the faster release of core materials such as probiotic cells (Zaeim et al. 2017). Therefore, to improve alginate properties when used in the pure form, it has been proposed to combine alginate with additional layers using different types of biopolymers (Koh et al. 2022). For example, chitosan – one of the most common coating materials – when used to coat Ca-alginate microcapsules prepared by double-stage technique successfully blocked bacterial release and maintained the viability of probiotics inside their structure under simulated gastrointestinal conditions (Zaeim et al. 2017).

Successful methods to produce multilayer particles are available, including co-extrusion, double emulsion, layer-by-layer, fluidized bed coater, and electrospray coating. In view that the main drawback during the manufacture is related to the lack of control over the thickness of the coating layer, the layer-by-layer technique has been often applied, as it allows better control of the coating dimensions, which is highly dependent on the number of deposited layers (Talebian et al. 2022). The method consists of the sequential deposition of oppositely charged polyelectrolytes onto substrates, whose deposition is driven by electrostatic interactions (Thomas et al. 2014). Such a technique has the potential to not only deliver probiotics but also improve the adhesion ability of probiotics to the intestinal mucosa (S. Li et al. 2023).

In this chapter, we will be discussed the main advantages of producing multilayer microparticles, have a closer look at the most common techniques used for coating, the impact of using one or more additional layers on probiotic protection, the frequently applied materials, as well as the main applications such as the target delivery using coated particles.

## Techniques

### Layer-by-Layer

The layer-by-layer method explores the deposition of thin films on solid supports, based on the electrostatic attraction between compounds with opposite charges. The first report in the literature emerged in the 1990s and showed that multilayer films composed of at least 100 alternating layers could be constructed, where the multilayer films were obtained by cyclically alternating the immersion of solid support in anionic and cationic solutions (Decher, Hong, and Schmitt 1992; Lvov, Decher, and Moehwald 1993)

Since its emergence, the technology of coating particles with biopolymers layers, known as layer-by-layer, has been one of the most explored for the protection of bioactive, as it allows the construction of thin film layers over particles in a more controlled, scalable, and economical way. One of the main functions of coating particles is to protect and control the permeability of the larger encapsulated components in addition to sustaining the permeation of small external molecules (Călinoiu et al. 2019; Zhao et al. 2019).

The layer-by-layer technique has been widely applied in the encapsulation of probiotics because, in addition to offering protection, it presents an extra advantage of regulating the functionality of the particles during the production, storage, and through the gastrointestinal tract, improving the probiotic viability and action (Călinoiu et al. 2019; González-Ferrero et al. 2020; W. Li et al. 2022). This technique still presents good versatility through a wide possibility of polymers that can be combined (e. g. alginate, chitosan, gelatin, pectin, starch, carrageenan, proteins, and lipids), as long as it is taken into account the adhesion of the layers while choosing the polymers (W. Li et al. 2022).

Given the diversity of polymers and encapsulated probiotic strains, some examples will be presented and discussed. Cheng and co-workers (2021) produced multilayer particles with alginate and protamine driven by enzymes (trypsin in the intestine). Among the formulations, the particles produced with two protamine layers balanced the disintegration, protecting the probiotics from gastric acids and bile, enabling bacterial adhesion, and successful colonization in the colon and ileum during *in vivo* testing.

*L. casei* was encapsulated in a double emulsion and coated with alginate and chitosan by electrostatic deposition. The resistance of *L. casei* cells to gastrointestinal conditions improved considerably with the increase in the number of layers, showing a reduction of 7.3, 5.5, and 3.1 log CFU/g for the multiple emulsion particles, with alginate coating layer and a second layer of chitosan, respectively (Beldarrain-Iznaga et al. 2021). In another study, green tea polysaccharides (gTPC) and gelatin were used as encapsulation materials for the protection of *L. acidophilus* using the layer-by-layer technique. The results showed that encapsulation was more complete with an increasing number of layers (1, 2, 4, and 6 polymers layers). Survival of non-encapsulated cells after the simulated gastrointestinal tract was 53.26% and encapsulated cells with different deposited layers (1 to 6) were 66.00%, 71.99%, 76.94%, and 79.89%, respectively (X. Chen et al. 2023).

Figure 1 shows the scanning electron microscopy (SEM) of particles produced by layer-by-layer assembly. The core was produced by dripping the emulsion-filled hydrogel in a calcium bath (surface charge ~ -60 mV before $CaCl_2$ gelification process) and was subsequently covered in a double layer with gelatin and alginate (~ +20 mV and -70 mV, respectively) (Silva et al. 2020). The particle was produced on a millimeter scale, enabling the fracture to visualize the internal structure and the coating layer (Figure 1C).

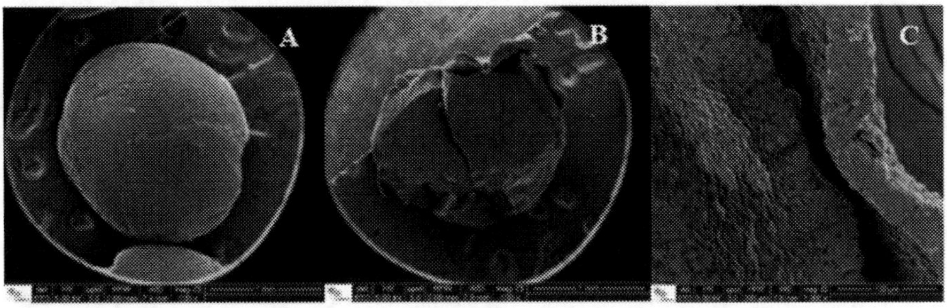


**Figure 1.** Emulsion-filled hydrogel produced using the layer-by-layer technique. Core composed of sunflower oil, alginate, gelatin, and gelatinized potato starch. The first layer was composed of gelatin at pH 3 and the second layer of alginate. a) External structure, b) Cross-sectional view of hydrogel particles, c) coating layer. Magnification of 100× (a, b) and 1000× (c). Scale bar: 1 mm (a, b) and 100 μm (c).

## Fluidized Bed Coater

For the production of coated particles using a fluidized bed, an ascending fluid flows through a column at sufficient speed to support the particles to be coated, without carrying them. This system resembles a boiling liquid, hence the term fluidization. For the production of coated particles, during fluidization, the coating fluid and a gaseous phase are in contact with the core, to promote their coating and drying (Dewettinck and Huyghebaert 1999; Zaghari, Basiri, and Rahimi 2020). This technique was developed in 1950 by Würster (Arshady 1993), presenting benefits due to the facilitated interactions between particles and the low temperatures used.

Different configurations can be obtained in fluidized beds, being the top-spray method the most used. For the production of encapsulation systems by the aforementioned method, the coating liquid is atomized in countercurrent over the fluidized particles. Throughout the entire cycle of the process, the particles receive coating uniformly due to fluidization and are subsequently collected. The method allows the coating of particles of different sizes, such as in the order of 2 to 100 μm. In the bottom-spray method, the most notable difference is in the short distance traveled by the coating material, which practically does not evaporate, forming films with lower viscosity and denser, with superior physical quality. Other configurations allow for tangential, continuous, or rotating fluidized bed coating (Dewettinck and Huyghebaert 1999). Particularly, each technique offers its advantages and disadvantages and must be evaluated concerning scaling, cost, coating effectiveness, and particle morphology, among other characteristics of interest.

A fluidized bed was used to perform the double coating of *Lactobacillus reuteri*, the particles were evaluated regarding the probiotics resistance to drastic temperature conditions (80 °C for 15 and 30 minutes) and pH (gastric simulation). Different concentrations of shellac and sodium alginate were evaluated in the first layer, and microcapsules coated with 1% sodium alginate showed greater survival after 1h in gastric simulation. Chitosan and gum Arabic were tested in the second layer. The results showed that the relative survival of bacteria in microcapsules with alginate and chitosan was higher for acidity and heating (Zaghari, Basiri, and Rahimi 2020). Chitosan is capable of adhering to the alginate surface, contributing to the reduction of the pore, and reinforcing the probiotic protection.

Similar research was carried out by Mirzamani et al. (2021) to improve the survival of *Lactobacillus acidophilus* encapsulated in a double layer using a fluidized bed. This work explored the application of probiotics in bread, evaluating the resistance of probiotics during bread baking and against simulated gastrointestinal conditions. Xanthan and alginate were used in the first coating layer, the second layer was produced with chitosan and gellan. The results showed that the particles produced with 1% alginate and 0.5% chitosan resulted in greater probiotic viability (24.3%) when subjected to heat treatment, evidencing the positive effect of encapsulation.

Schell and Beermann (2014) used the technique of granulation and coating with whey powder and shellac in a fluidized bed with a focus on improving *L. reuteri* viability. Surprisingly, bacterial counts after 28 days of storage at 4°C were the same for coated and uncoated cells. However, encapsulated probiotics showed a higher survival rate (76.74%) compared to free cells (17.74%) after gastric intestinal simulation. In another study, among different materials tested, whole milk powder matrix showed better coverage of *Lactobacillus paracasei* by fluidized bed, the viability in 4 weeks of storage at 25°C was approximately 9 log CFU/g, against an initial viability count of 9.51 log CFU/g (Poddar et al. 2022).

## Electrospray Coating

The technique of producing particles by electrospray consists of atomizing a polymeric solution with a peristaltic pump coupled to a capillary nozzle maintained at a high electric potential. Drops or jets are formed and submitted to electrical shear stress towards a solid or liquid collector, therefore, dry or wet encapsulation systems can be obtained, respectively. Once a droplet or jet is formed and exposed to an electrical field, repulsive forces overcome surface tension and viscoelastic forces, and the formed jet may (drip mode) or not be broken (jet mode) into smaller droplets while directed to the collector (Jaworek, Sobczyk, and Krupa 2018; Zaeim et al. 2017).

As with other encapsulation techniques, the process conditions directly interfere with the characteristics of the encapsulation systems obtained such as the type of biopolymers, concentration, feed flow rate, operating voltage, distance between the needle tip and the collector solution, configuration of electrospray nozzles and deposition times for electrospray coating. For example, the configuration of the nozzles used in electrospraying can allow the flow of different polymers simultaneously, in concentric tips. This technique can be used for coated particle production (Ghorani and Tucker 2015; Jaworek, Sobczyk, and Krupa 2018).

Among the advantages of the electrospray method for encapsulation, it can be mentioned the obtaining of small particles with low polydispersion, the possibility of scaling up for industrial use, and the employment of solutions with high viscosity and mild processing conditions, which has aroused interest in probiotic encapsulation due to their sensitivity to heat, besides the efficiency of material deposition that allows the production of coating layers (Almería, Fahmy, and Gomez 2011; Ghorani and Tucker 2015; Librán, Castro, and Lagaron 2017).

Zaeim and co-authors (2017) produced microcapsules of alginate and chitosan by electrospray to protect *Lb. plantarum* cells. The authors investigated the production of single-stage or double-stage encapsulation systems, showing that *L. plantarum* encapsulated by single-stage had low viability and protection against gastric simulation, which was attributed to the antimicrobial activity of chitosan. On the other hand, the alginate microcapsules coated with chitosan, by the double-stage procedure, improved probiotics protection and viability during simulated gastric conditions in 1 log cycle compared to the free cells, attributed to delayed diffusion of $H^+$ ions by external layer.

A new encapsulation method using electrospray was presented by Librán, Castro, and Lagaron (2017). The results showed the potential to increase the stability of powders, namely lyophilized *bifidobacterial*, with coating by electrospray using different hydrocolloids. Electrospray coating with whey protein concentrate showed the best survival results of probiotics under storage conditions (23% RH, 23°C, and 37°C). The *bifidobacteria* viability of the electrospray-coated samples was 6 log CFU/g after 600 days at 23 °C and approximately 6 log CFU/g after 10 days at 37 °C, while the control samples completely lost viability in the same period.

The electrospray technique was also used to produce microcapsules containing *L. plantarum*. Alginate and resistant starch-alginate capsules, coated and uncoated with chitosan, were produced. Electrospray significantly enhanced the survival rate of the probiotic bacteria during simulated gastric conditions. For free *L. plantarum*, after 1 hour of incubation, no viable cells were found. Starch-alginate microcapsules promoted greater viability compared to chitosan-coated alginate particles (reduction of 2.14 log CFU/mL and 3.68 logs CFU/mL after 2h), which was attributed to the starch prebiotic action. However, the coating provided greater gastrointestinal mucosa adhesion (Phuong Ta et al. 2021).

The literature presents multiple studies involving the electrospray technique for the encapsulation of probiotics, covering the application of different cells and wall materials applied. For example, works are investigating electrospray alginate microcapsules coated with citric acid-modified zein for *L. acidophilus* protection (Laelorspoen et al. 2014), a combination of encapsulation materials (alginate and pectin) to improve the viability of *L. plantarum* upon storage and during simulated passage through gastrointestinal fluids (Coghetto et al. 2016), a comparison of methods and polymers for encapsulation of *L. rhamnosus*, such as electrospraying, freeze, and spray drying (Moayyedi et al. 2018), in addition to combined techniques, namely electrospraying and fluidized bed coating, to obtain multilayer systems for the protection of *L. acidophilus* (Pitigraisorn et al. 2017).

These results highlight the potential of the electrospray technique to protect probiotics, proving to be a promising, low-cost, and scalable encapsulation technology that can operate with various biopolymers. Figure 2 presents a schematic representation of the processes addressed in this chapter.

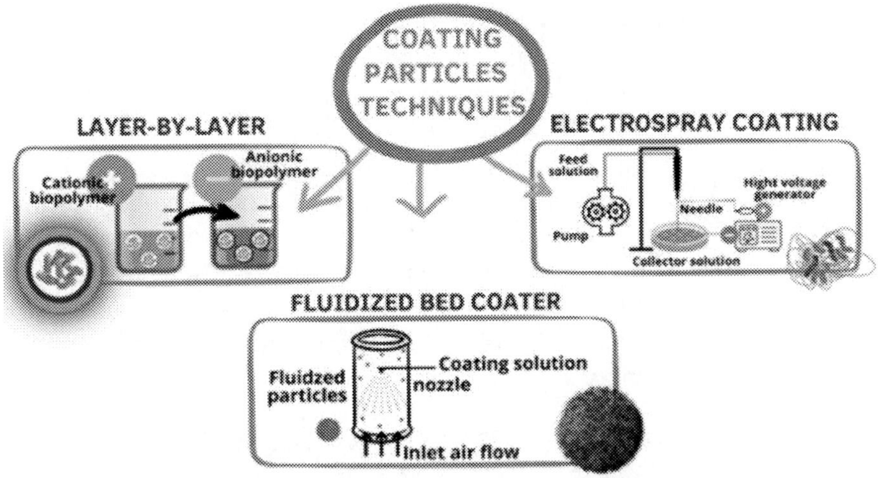


**Figure 2.** Schematic representation of different techniques for producing multilayer particles.

## Morphology of Multilayer Particles

Particle size is a characteristic that interferes with the protection of microorganisms, the texture, and the sensory characteristics of the products to which they are added. In this manner, to ensure the protection of probiotics, it is known that particles with sizes larger than probiotics must be produced. For example, *Lactobacillus* have the shape of rods with a length that can vary between ~1 and 1.5 μm and a diameter of ~0.7 to 1 μm (Schär-Zammaretti and Ubbink 2003). According to Călinoiu et al. (2019), particles with sizes smaller than 100 μm have better sensory properties, in addition to interfering less with rheological properties. With the technique of spray-drying small particles that guarantee protection of probiotics can be obtained. Particles produced by electrospray coating can vary in size (Librán, Castro, and Lagaron 2017), however, the average diameter is approximately 2 μm.

In general, particles formed by ionic gelation produced by dripping or atomization have a spherical shape, as evidenced in Figure 3. In this case, it is possible to observe part of the probiotics on the surface, characterizing it as a microsphere and not as a microcapsule, where the core is completely covered by the wall material.

Hills et al. (2000) observed the formation of a uniform and spherical alginate matrix with small pores by dripping. The authors verified that the internal structure varies according to the concentration of salts and alginate. According to Santos and Machado (2021), this typical alginate porosity can compromise the probiotic viability, reinforcing the potential of coating layers. Particles obtained by double-layer co-encapsulation of probiotics were produced by Zaeim et al. (2019). For this, inulin and resistant starch were added to the alginate network and electrospray into $CaCl_2$ solution, which resulted in increased particle size and reduced wrinkles on the surface. The coating layers were obtained by immersing the particles in a chitosan solution to overcome permeability due to porosity.

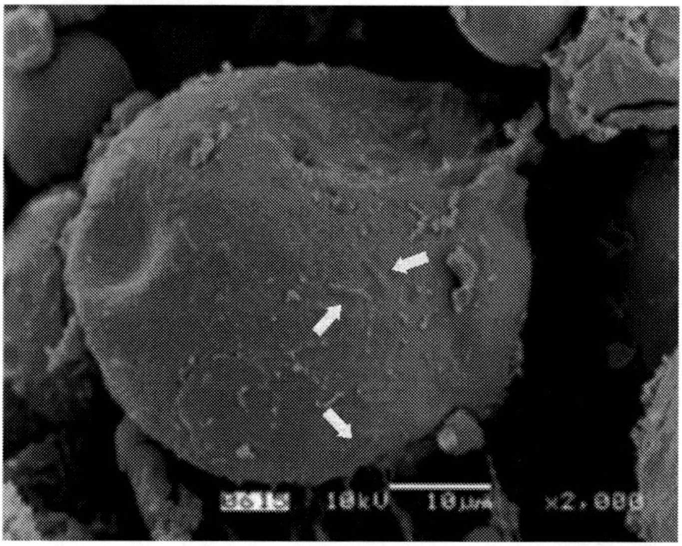


**Figure 3.** Spherical structure *L. acidophilus* particle produced by ionic gelification using an atomizer. Note that it is possible to observe some *L. acidophilus* on the surface particle (white arrow).

Depypere et al. (2009) quantified the thickness and quality of the coating of microparticles with protein (gelatin and sodium caseinate) produced in a fluidized bed. They obtained perfectly spherical particles, however, process conditions such as the distance between the nozzle and the core particle bed, besides the coating solution concentration affected the quality of the coating layer.

Particles for the protection of *L. casei* were obtained by double multilayer emulsion, using alginate and chitosan, all of which had a semi-spherical shape.

The differences were attributed to the materials and number of layers. When modified with chitosan by layer-by-layer, these particles became more spherical, rough, and less porous (Beldarrain-Iznaga et al. 2021). Single and double emulsion particles presented an irregular surface, and the adhesion of the alginate layer promoted the formation of a smoother and more porous surface (Santos and Machado 2021; Zaeim et al. 2019).

The application of the fluidized bed technique for coating particles aims to obtain a more uniform morphology. In general, the formation of complete coating coverage occurs after many pass through the fluidized bed. The type of nozzle affects the size and distribution of drops since pneumatic nozzles are more controllable compared to hydraulic ones. Another condition that determines the morphology is the air used for atomization, which causes evaporation and increases droplet viscosity, thus inhibiting scattering and coalescence upon contact with the core material (Dewettinck and Huyghebaert 1999).

The electrospray technique can form systems with varied morphologies, from capsules to fibers. These morphologies specificities can be achieved through adjustments in processing parameters (applied potential, electric field, distance, flow rate) and the materials used (conductivity, viscosity, surface tension, dielectric constant), in addition to other forms such as tubes, spirals, rings, and other shapes that can range from nano to microscale (Mendes and Chronakis 2021).

For example, by alternating materials, Librán, Castro, and Lagaron (2017) obtained spherical particles and beaded fibers to protect probiotics by electrospray coating atomization in a three-step process. Particles were obtained using whey protein, commercial-resistant maltodextrin, maltodextrin, and zein. Beaded fibers were obtained using plastic derived polymer polyvinylpyrrolidone (PVP). The particles were formed by jet breakage, where beaded fibers resulted from the use of solutions with higher viscosity or low surface tension.

Probiotic bacteria (*Lactobacillus reuteri*) with single coating by fluidized bed produced with shellac and sodium alginate showed porosity and oval shape. With the addition of a second layer of Arabic gum by the fluidized bed coating, cracks in the microcapsules were observed, while the samples coated with chitosan showed more uniformity on external surfaces (Zaghari, Basiri, and Rahimi 2020). Similarly, particles with a double layer of alginate and gellan were also produced by the fluidized bed to protect *L. acidophilus*. Alginate single-layer particles did not show regular size and shape, and the presence of pores was attributed to the drying process. With the deposition of the second layer, the pores were reduced and a smoother appearance was observed (Mirzamani et al. 2021).

Coating quality in fluidized beds was evaluated at both top and bottom spray settings. Glass microspheres with an average particle diameter of 204 μm were used for coating with sodium caseinate and gelatin. The particle size of the samples did not differ significantly between experiments. However, it has been found that coating quality increases significantly with higher coating levels. For all experiments, however, problems with coating the microparticles were encountered. The higher the viscosity of solutions and the greater the distance traveled by the coating material until reaching the core, reflected in decreased coating quality (Depypere et al. 2009).

Faced with so many possibilities of processes and materials used in the multilayer encapsulation systems, varied morphological results can be obtained, since the interaction of materials and preparation conditions will result in systems with porosity, shape, and surfaces that can meet the specificities for each desired application.

## Multilayer Particles: Coating Composition

In the last years, microencapsulation in multilayer particles has been more commonly used to improve the stability of encapsulated compounds against processing, digestive and storage conditions (de Araújo Etchepare et al. 2020). Different materials can be exploited to coat particles including gelatin (Kerdsup et al. 2022), whey protein (Bakhtiyari, Hamidi-Esfahani, and Barzegar 2022), besides alginate (Beldarrain-Iznaga et al. 2021), gellan (Mirzamani et al. 2021), pectin (Menezes et al. 2019) and chitosan (Masoomi Dezfooli et al. 2022) – the most applied material to cover alginate microgels. Table 1 shows an overview of the biopolymers used to cover particles designed to encapsulate probiotics. In general, it is noticed that alginate microgels are often used to build the core of the particle (Table 1). This is attributed to the ease of handling, non-toxic nature, mucoadhesive properties, and low cost of alginate, besides its ability to increase the viability of probiotic bacteria when exposed to different conditions compared with free cells (Etchepare et al. 2015). Despite the suitability of alginate microcapsules for encapsulation, the particles formed usually present high porosity and susceptibility to disintegration at acidic media and in the presence of excess monovalent ions, and calcium-chelating agents such as phosphate and citrate (Khosravi Zanjani et al. 2014). To overcome this hurdle while decreasing the loss of encapsulated material and improving stability of probiotics, different approaches have been proposed such as addition of fillers (Q. Li et al. 2023), co-encapsulation with prebiotics (Sathyabama et al. 2014), and coating alginate microparticles with biopolymers through electrostatic interactions (Bakhtiyari, Hamidi-Esfahani, and Barzegar 2022).

Numerous works have shown the advantages of using chitosan as a coating agent to improve the properties of alginate particles (Masoomi Dezfooli et al. 2022; Bakhtiyari, Hamidi-Esfahani, and Barzegar 2022; Iqbal et al. 2022). This combination is interesting to overcome the limitations of each material since alginate generally shrinks at low pH and dissolves at high pH values. Oppositely, chitosan solubilizes in acidic media, while becoming insoluble in the basic environment. This mixture is known for being able to prolong the contact time of active ingredients with epithelium, besides enhancing the absorption by opening the intracellular tight junctions (Niculescu and Grumezescu 2022).

Chitosan as a coating material has been exploited in multiple ways. For example, a previous study has shown the benefit of using chitosan as a coating material exploring different processes of chitosan incorporation – single or double stage. In the first case, microcapsules are obtained in a solution containing the crosslinking agent ($Ca^{2+}$) and chitosan and, in the second case, the coating is performed only after gelation is carried out (Zaeim et al. 2017). Authors demonstrated that particles prepared by the double-stage procedure presented improved viability under simulated gastrointestinal conditions when compared to free cells and probiotics encapsulated by single-stage. Furthermore, another study has also highlighted the good coating properties of chitosan to protect three different probiotic bacteria and decrease the release of the encapsulated probiotics in seawater. The coated-chitosan beads allowed a higher probiotic load in the digestive tract of fish fed with these beads compared to non-coated particles (Masoomi Dezfooli et al. 2022).

Although for most cases coated particles can increase the stability and protect probiotics, it has been shown that sometimes modifying the surface does not enhance delivery, as reported for chitosan-coated alginate microgels used to protect *Bifidobacterium longum* cells (Yeung et al. 2016). It is possible that the bacteria strain can influence the protection efficiency provided by the particles, as some resistant bacteria would not require as much protection. Indeed, the study highlighted that encapsulated bifidobacterial viability is dependent on subspecies as well as strains, pointing out a range of genotypic and phenotypic factors that influences stress responses related to probiotic viability (Yeung et al. 2016).

Protein coating has also been studied to increase the stability of probiotics (Table 1). Although it has been reported that multilayer particles solely based on proteins are not successful to protect *Lactobacillus rhamnosus* when encapsulated in alginate microbeads (Yucel Falco et al. 2019), others have suggested that protein coating as those prepared with whey protein isolate can be more effective at protecting *Lactobacillus plantarum* cells against simulated gastrointestinal fluids than coating with chitosan (Bakhtiyari, Hamidi-Esfahani, and Barzegar 2022).

Multilayer particles coated with the same biopolymer can also be applied to protect probiotic cells. A study has shown the ability of gum arabic and β-cyclodextrin (9:1) besides hydrogenated palm oil to be used as wall material when choosing either spray drying or spray chilling to form single or double-layer microparticles, aiming the protection of different microorganisms (Arslan-Tontul and Erbas 2017). The authors demonstrated that the probiotic counts of single-layered microcapsules exposed to heat were generally higher than that of double layered microcapsules. However, *L. acidophilus* survived better when heat exposed in double layered microcapsules. Similarly, Mokarram et al. (2009) showed that among the different capsules prepared with alginate (uncoated, single coated, and double coated), double coating provided the best protection to cells. Although sometimes double coatings provide extra protection to the cells when encapsulated within layers of the same polymer, one should keep in mind that the type of cell can highly impact the outcome.

In addition, a great number of works in the literature have evaluated the impact of adding layers on probiotic cell viability. For instance, it has been reported recently that double-coated beads were the most effective to protect *Lactobacillus plantarum* isolated from fermented tea leaves and encapsulated in multilayer beads. Encapsulation in an alginate bead coated with a single (ACG) or double layer (ACGD) of gelatin prolonged the survival of *L. plantarum* TISTR 2074 in making juice for 3 and 4 days, respectively, while the viability of the probiotic *L. plantarum* MF 105, on the other hand, was extended for more than 5 and 7 days in ACG and ACGD microcapsules, respectively (Kerdsup et al. 2022).

**Table 1.** Multilayer particles are produced by different biopolymers and methods of fabrication

| Inner Material | Coating Material | Probiotic | Results | Reference |
|---|---|---|---|---|
| Alginate | Alginate + single-layer of gelatin 0.2% (ACG) and Alginate + double-layer of gelatin 0.2% (ACGD). | *L. plantarum* MF105 and TISTR 2074 | ACGD beads were the most effective in cell protection. Cell adaptation in combination with encapsulation enhanced the survival of probiotic cells in sub-lethal stress conditions. Data had shown that increasing the coating layer provided a chance to improve the lethality of probiotic bacteria. | (Kerdsup et al. 2022) |
| Alginate | Chitosan 0.2, 0.4, and 0.8% w/v. | *Exiguobacterium* sp., *Vibrio* sp., and *Enterococcus* sp. | Chitosan coating improved the stability of beads and reduced the release of the encapsulated probiotics in seawater. A higher probiotic load was found in the digestive tract of animals fed with chitosan-coated alginate beads compared to the control group. | (Masoomi Dezfooli et al. 2022) |
| Alginate | Whey protein (1.38 and 2.76%) or chitosan (0.5 and 0.7 g/100 mL). | *Lactobacillus plantarum (MT, ZH593)* | The release of extract from alginate beads was retarded by coating them with whey protein. Microbeads coated with whey protein isolate (2.76%) resulted in higher protection in simulated gastrointestinal fluids than coating with chitosan. | (Bakhtiyari, Hamidi-Esfahani, and Barzegar 2022) |
| Alginate | Single coating with sodium alginate (SA) at 4%, 5%, and 6% and chitosan (CH) at 0.6%, 0.8%, and 1.0% concentrations. Whey protein concentrate (WPC) at concentrations of 5.0%, 7.5%, and 10% was used for double coating of SA 6% and CH 0.8% microbeads. | *Bifidobacterium bifidum* | The encapsulation efficiency of double-coated microbeads was greater than 90% as compared to single coated microparticles. Double coating with the polysaccharides–protein matrix maintained viability (>$10^6$ CFU/ml) for 28 days with no prominent change in physicochemical and sensory attributes as compared to free cells. | (Iqbal et al. 2022) |

**Table 1.** (Continued)

| Inner Material | Coating Material | Probiotic | Results | Reference |
|---|---|---|---|---|
| Alginate | Formulations:<br>A1 - Alginate microparticles with a layer of WPC 4%<br>A2 - Alginate microparticles with a layer of WPC 4% and a layer of alginate 0.125% A3 - Alginate microparticles with a layer of WPC 4%, a layer of alginate 0.125% and a layer of WPC 2.5%. | *Lactobacillus acidophilus* | The encapsulation efficiency obtained was higher than 80% in all multilayer treatments. The addition of multilayers and/or materials of a different chemical nature may provide better thermotolerance to microencapsulated probiotics, since all multilayer treatments tolerated better thermal treatments, mainly at temperatures of 63 °C for 30, 60, and 120 min. | (de Araújo Etchepare et al. 2020) |
| Pectin | Whey protein (2% and 0.5%) and pectin (2% and 0.3%). | *Lactobacillus acidophilus* | When heat treated, microcapsules with a layer of whey protein (LA/P1) maintained the viability of their encapsulated *Lactobacillus acidophilus* (9.57 log CFU/g). The best storage viability was at -18 °C, with a count of 7.86 log CFU/g at 120 days for microcapsule LA/P1. Microparticles consisting of two layers of whey protein (LA/P3) presented a 6.55 log CFU/g at 105 days. | (Menezes et al. 2019) |
| Alginate | Single stage: microcapsules have been obtained in a solution containing crosslinking agent ($CaCl_2$) and chitosan. | *Lactobacillus plantarum* | Encapsulated *Lactobacillus plantarum* using the double-stage procedure improved the viability under simulated gastrointestinal conditions in comparison to free cells and to cells encapsulated by the single-stage procedure. For the latter, it was found that too close proximity of the chitosan layer to the protected cells harms cell viability. That effect was | (Zaeim et al. 2017) |

| Inner Material | Coating Material | Probiotic | Results | Reference |
| --- | --- | --- | --- | --- |
| | Double-stage: gelation process was carried out first and the microcapsules have been coated with chitosan subsequently. | | explained by the increased antimicrobial activity of chitosan resulting from the presence of protonated amine groups in chitosan under acidic conditions. | |
| Alginate | Hybrid - Chitosan, dextran sulfate, lactoferrin, polymerized peptides, sodium caseinate, whey protein isolate, alginate. | *Lactobacillus rhamnosus* GG | Multilayers solely based on proteins are not successful and chitosan is required as a first stabilizing layer in the coating formulation. Protein-coated microbeads were partially dismantled, whereas polymer-coated microbeads were intact after a sequential treatment in simulated gastric and intestinal fluids. This suggests that hybrid formulation offers an advantage over the coatings based on biopolymer multilayers in terms of better release of bacteria. | (Yucel Falco et al. 2019) |
| Double emulsions - simple emulsion (Lc/O) was mixed with an outer aqueous dispersion of sodium caseinate (Cas) (17 g/L) to obtain the double emulsion (Lc/O/Cas) | Alginate and chitosan layers were formed using layer by layer technique. | *Lactobacillus casei* | The alginate-chitosan layers provided additional protection to *L. casei* cell membranes, substantially preserving the cholesterol assimilation ability, surface hydrophobicity, auto-aggregation, and co-aggregation of *L. casei* after simulated *in vitro* digestion. | (Beldarrain-Iznaga et al. 2021) |

**Table 1.** (Continued)

| Inner Material | Coating Material | Probiotic | Results | Reference |
|---|---|---|---|---|
| Tara gum | Tara gum (0.1%) + three species of amino acids (each 0.1%) + rice protein powder (1.0%) + red ginseng dietary fiber (1.0%). | *Lactobacillus (Limosilactobacillus) reuteri* KCTC 3594, *Lactobacillus (Lactiplantibacillus) plantarum* KCTC 3108, *Lactobacillus (Lacticaseibacillus) rhamnosus* KCTC 3237, *Lactobacillus (Lacticaseibacillus) casei* KCTC 3109, *Lactobacillus (Limosilactobacillus) fermentum* KCTC 3112 | Multilayer-coated probiotic strains exhibited higher survival rates under low pH conditions (maximum survival: 81.8% for *Lactobacillus casei*) and bile acid conditions (>50% survival for *Lactobacillus rhamnosus*) and exhibited significantly improved storage stability under various temperatures after freeze-drying when compared to uncoated control strains. Multicoating also increased the ability of probiotic strains to adhere to intestinal epithelial cells. For example, encapsulated *L. reuteri* exhibited 56.9% adhesion, which was approximately four-fold higher than that of the uncoated strain. | (Jeon et al. 2023) |
| - | Whey protein isolate fibrils, (WPIFs), sodium alginate (ALG), carboxymethyl cellulose (CMC), and xanthan gum (XG). | *Lactobacillus plantarum* | The LbL coating with WPIF and CMC, ALG, and XG contributed to the strong protection of LP90 against SGFs. Electrostatic interactions and hydrogen bonding were involved in the LbL coating process. | (S. Li et al. 2023) |
| Alginate or Xanthan | Chitosan (0.5, 1, and 1.5% w/v), and gellan (0.5, 1, and 1.5% w/v) were used for the second layer. | *Lactobacillus acidophilus* | Alginate in the wall matrix made significantly higher viability of the probiotic against acid resistance. Moreover, 0.5% chitosan in the outer layer of the microcapsules caused the probiotic more resistant to the heat treatment of 90 °C for 15 min. | (Mirzamani et al. 2021) |
| Arabic gum and b-cyclodextrin (9:1) Hydrogenated palm oil | Arabic gum and b-cyclodextrin (9:1) Hydrogenated palm oil | *Saccharomyces boulardii, Lactobacillus acidophilus,* and *Bifidobacterium bifidum* | Double-layered microencapsulation improved heat and gastric resistance of *L. acidophilus* but had a similar survivability of *B. bifidum* cells to single-layered microcapsules, contrary to *S. boulardii* and *L. acidophilus* results. Spray drying ensured similar survivability with spray chilling, however, it had high protection in the simulated intestinal system and heat tolerance test. | (Arslan-Tontul and Erbas 2017) |

Particles coated with up to three levels of multilayer were also investigated by another study, aiming to increase the viability of *Lactobacillus acidophilus* under different storage conditions such as room, refrigeration, and freezing temperatures (de Araújo Etchepare et al. 2020). Results indicated that the best protection provided by the different coated particles varies with the storage times and temperatures. Interesting, under refrigeration temperature (7 °C), the microparticles without coating and containing double or triple layers presented no viability, while the treatment containing only one single layer of whey protein obtained still high counts (7.23 log CFU · $g^{-1}$). Conversely, the three layers of coating showed the least viability losses at freezing temperature (−18 °C) over 120 days of storage (reduction of 0.19 log).

Therefore, these works suggest that the desired number of layers to coat particles is related to different parameters such as the bacteria strain, the conditions the probiotics are submitted to, and the time frame of evaluation. Moreover, it should be carefully considered that increasing the number of coating layers does not always increase probiotic viability.

Finally, multicoating can also increase the ability of some probiotic strains to adhere to intestinal epithelial cells. For example, besides increasing the viabilities of the probiotic strains to 77.15%, 81.83%, 79.62%, and 79.51% in *L. reuteri, L. casei, L. plantarum, and L. rhamnosus*, respectively, the four-layer coatings of 0.1% tara gum, 0.1% amino acids, 1.0% rice protein powder, and 1.0% red ginseng dietary fiber also improved the adhesion abilities of *L. Reuteri* and *L. Rhamnosus* up to 56.91% and 46.22%, respectively. In contrast, the adhesion abilities of *L. casei* and *L. Plantarum* were not improved by multilayer coating (Jeon et al. 2023).

## Environment-Adjustable Release: Designing Target Delivery with Coated Particles

The literature is well elucidated as to the knowledge that in acidic pH, alginate does not present good protection capacity due to its porosity and affinity for ions and chelators (Zaeim et al. 2017). This fact reinforces that the physical stability of encapsulation systems needs to be controlled by combining alginate with other materials or by adding additional layers.

In the study carried out by Yucel Falco et al. (2019), alginate microspheres coated with multiple layers of biopolymers and proteins were tested under simulated acidic stomach conditions and compared to the protective performance of uncoated alginate microspheres containing probiotics. Chitosan was required as the first layer to reduce permeability. Polymer and protein-based coatings proved to be a good combination, offering high resistance against enzymatic hydrolysis. The more complex adsorption of proteins, due to their compact structure and non-uniform charge distribution benefited by the addition of polymers. Considering they present a more flexible structure and uniform charges, these polymers worked as a binding agent, ensuring greater *L. rhamnosus* viability.

Several studies have demonstrated the additional protection of probiotic particles coated with chitosan such as chitosan-coated agar-gelatin particles for *L. plantarum* protection (Albadran et al. 2020) and chitosan-coated alginate particles to improve stability of *S. boulardii*, *B. longum* and *L. rhamnosus* (Ji et al. 2019; Qi et al. 2020; Santos and Machado 2021). The protective action of chitosan in acidic conditions can be attributed to its buffering capacity and to retard the acid diffusion in encapsulation systems, in addition to reducing porosity. It is also noted that the thickness, concentration, and composition of the coating material influence the release of cells at a pH close to neutrality (Cook et al. 2013).

Aiming the protection of Bifidobacterium, particles of xanthan coated with chitosan and covered with a second xanthan layer were produced by the extrusion technique (Chen et al. 2017). Xanthan and chitosan formed cross-link gels with reversible ionic bonds, offering high resistance to pH and enzymatic variations. Greater viability during the simulation of gastric fluids was achieved for double-coating particles and attributed to the buffering capacity of the outer layer, which in acidic conditions, reduces electrostatic interactions and promotes the absorption of water to buffer the acidic compounds. The double-layer particles also stood out when subjected to bile salts, considering the ion exchange reduces porosity and, consequently, the stress conditions. After 20 min in contact with intestinal fluids, controlled delivery was achieved, however, the single layer showed the best release profile.

Figure 4 illustrates the structure of a particle produced from alginate that received gelatin and alginate as coating layers. It is possible to observe the greater porosity of the uncoated particle and a more intact internal structure for the coated one after the gastric simulation. Because the interface of the uncoated particle is more exposed to the action of gastrointestinal fluids, the weakening of the gel network at the edge allows easier action of enzymes and salts. This leads to greater destabilization compared to the coated particle, even if the covering layer is composed of alginate (Silva et al. 2020).

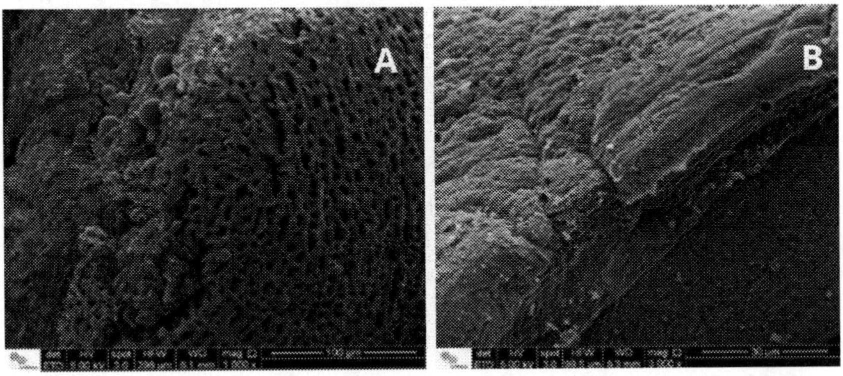


**Figure 4.** Structure of particles after *in vitro* gastrointestinal simulation. The outer layer of particles is produced from potato starch, alginate, gelatin, and sunflower oil without coating (a). The external layer of the same previous particle coated with gelatin and alginate (b).

## Probiotic Protection Performance of Coated Particles

Nowadays, the incidence of gastrointestinal diseases has been increasing worldwide, which brings economic and social impacts on healthcare costs and impaired quality of life. In this context, probiotics have received great attention due to the benefit of alleviation of gastrointestinal conditions. They are not only capable of reestablishing gut microbiota homeostasis and restoring gut barrier integrity, but also they can protect against invading pathogens and prevent chronic inflammation (Talebian et al. 2022). To exert these activities, however, it is important that the cells arrive at the active site and thus, they have to stand harsh storage conditions and survive the gastric and duodenal transit, ensuring they are viable and in sufficient numbers to provide the expected benefits (Mogna and Mogna 2014).

In that regard, the main applications of multilayer microencapsulated probiotics have often been related to the improvement of thermal stability and storage conditions, besides ensuring survivability in gastric and intestinal environments (de Araújo Etchepare et al. 2020; Mirzamani et al. 2021). Previous work has shown that coating alginate microparticles with whey protein (WPI) rather than chitosan created a good matrix to protect and increase *L. plantarum* viability in the lower part of the gastrointestinal tract, besides allowing the controlled delivery of *Silybum marianum* seed extract (a non-carbohydrate prebiotic) (Bakhtiyari, Hamidi-Esfahani, and Barzegar 2022). Authors associated the better probiotic protection provided by WPI with the acidic endurance and high buffering capacity of this biopolymer.

In addition, a prior study has developed multilayer (up to 10 layers) coated lactic bacteria or *bifidobacteria* with probiotic activity to prepare finished products (e.g., foods, medical devices, supplements, or pharmaceutical formulas) that overcome major drawbacks found in the field. In other words, authors created lipid-based coatings for microencapsulating bacteria that do not exhibit or have less dependency on the pH of media or the chemical composition of the food; present the necessary resistance to pass through gastric and duodenal transit intact and show less aggregation and damage from mechanical stresses or friction created during the processing steps (Mogna and Mogna 2014). In summary, the success of the invention relied on the choice of lipids to create the coating layers that are dependent on the chemical and physical properties of the finished product, the water content, and the ingredients and additives used in the formulation.

The use of probiotics as an alternative to antibiotics in livestock production has also been suggested. Varied lactic acid bacteria (LAB) strains encapsulated in coated-chitosan alginate beads prepared by different methods (i.e., extrusion, emulsion, and spray drying methods) showed the highest viability of spray-dried particles after storage for 6 months at room temperature. Moreover, all encapsulation methods still preserved antibacterial activity, which is interesting for industrial processes and livestock farm usage (Pupa et al. 2021).

Finally, to improve probiotic administration in aquaculture minimizing environmental contamination and loss of bacterial viability to grow black-footed abalone (*Haliotis iris*), a study demonstrated the benefit of using chitosan-coated alginate beads that not only improved the stability of particles but also reduced the release of the encapsulated probiotics in seawater. According to the authors, the digestive tract of animals fed with these beads presented a higher probiotic count ($1.3 \times 10^8$ CFU) compared to the control group ($6.0 \times 10^3$ CFU) (Masoomi Dezfooli et al. 2022).

Although there is a growing interest in producing multilayer particles to enhance the survival of probiotics, it is still noticed a lack of studies on the application of these types of particles in finished products, and further research is still required to better unravel their impact when added to the final application.

## Conclusion

The great development that science has provided in the area of encapsulation is notorious. Currently, the variety of encapsulation systems, from the simplest - based on one or two coating steps, to the sophisticated systems - based on multilayers, in addition to the range of materials and processes, ranging from simple to complex equipment, allow the production of countless particles. In addition to improving protection, and ensuring controlled delivery, one of the main driving forces for research development focuses on using renewable, biocompatible, more environmentally friendly, and safe-to-consume materials. In this sense, it is worth mentioning the use of plant-based proteins to create microparticles as a future trend. Process scaling for industrial production and cost reduction should also be considered to increase the supply of encapsulated probiotic products in the food market.

## References

Albadran, Hanady A, Andrea Monteagudo-Mera, Vitaliy V Khutoryanskiy, and Dimitris Charalampopoulos. 2020. "Development of Chitosan-Coated Agar-Gelatin Particles for Probiotic Delivery and Targeted Release in the Gastrointestinal Tract." *Applied Microbiology and Biotechnology* 104 (13): 5749–57.https://doi.org/10.1007/s00253-020-10632-w

Almería, Begoña, Tarek M Fahmy, and Alessandro Gomez. 2011. "A Multiplexed Electrospray Process for Single-Step Synthesis of Stabilized Polymer Particles for Drug Delivery." *Journal of Controlled Release* 154 (2): 203–10. https://doi.org/10.1016/j.jconrel.2011.05.018

Araújo Etchepare, Mariana de, Graciele Lorenzoni Nunes, Bruna Righi Nicoloso, Juliano Smanioto Barin, Erico Marlon Moraes Flores, Renius de Oliveira Mello, and Cristiano Ragagnin de Menezes. 2020. "Improvement of the Viability of Encapsulated Probiotics Using Whey Proteins." *LWT* 117: 108601. https://doi.org/10.1016/j.lwt.2019.108601.

Arshady, R. 1993. "Microcapsules for Food." *Journal of Microencapsulation* 10 (4): 413–35. https://doi.org/10.3109/02652049309015320

Arslan-Tontul, Sultan, and Mustafa Erbas. 2017. "Single and Double Layered Microencapsulation of Probiotics by Spray Drying and Spray Chilling." *LWT - Food Science and Technology* 81: 160–69.https://doi.org/10.1016/j.lwt.2017.03.060.

Bakhtiyari, Maryam, Zohreh Hamidi-Esfahani, and Mohsen Barzegar. 2022. "Optimization of Co-Encapsulation of L. Plantarum Cells and Silybum Marianum Seed Extract and Evaluation of Protective Effect of Extract on Cells Survival in Simulated Gastrointestinal Fluids." *LWT* 165: 113733. ttps://doi.org/10.1016/j.lwt.2022.113733.

Beldarrain-Iznaga, Tatiana, Ricardo Villalobos-Carvajal, Eva Sevillano-Armesto, and Javier Leiva-Vega. 2021. "Functional Properties of Lactobacillus Casei C24 Improved by Microencapsulation Using Multilayer Double Emulsion." *Food Research International* 141: 110136. https://doi.org/10.1016/j.foodres.2021.110136.

Călinoiu, Lavinia-Florina, Bianca E Ştefănescu, Ioana D Pop, Leon Muntean, and Dan C Vodnar. 2019. "Chitosan Coating Applications in Probiotic Microencapsulation." *Coatings*. https://doi.org/10.3390/coatings9030194.

Cezarino, Eliza Cristina, Karen Cristina Guedes Silva, Flávia Souza Almeida, and Ana Carla Kawazoe Sato. 2022. "Stability and Viability of Synbiotic Microgels Incorporated into Liquid, Greek and Frozen Yogurts." *Journal of Food Science* 87 (4): 1796–1809. https://doi.org/10.1111/1750-3841.16107.

Chen, Li, Ting Yang, Yajuan Song, Guowei Shu, and He Chen. 2017. "Effect of Xanthan-Chitosan-Xanthan Double Layer Encapsulation on Survival of Bifidobacterium BB01 in Simulated Gastrointestinal Conditions, Bile Salt Solution and Yogurt." *LWT - Food Science and Technology* 81: 274–80. https://doi.org/10.1016/j.lwt.2017.04.005.

Chen, Xiaoqiang, Ke Zhang, Yan'an Wei, Yongyong Wang, Xiaofan Wu, and Haifeng Wan. 2023. "Green Tea Polysaccharide Conjugates and Gelatin Enhanced Viability of L. Acidophilus by Layer-by-Layer Encapsulation." *Food Bioscience* 52: 102471. https://doi.org/10.1016/j.fbio.2023.102471.

Cheng, Qikun, Lu Liu, Mingzhi Xie, Hang Li, Dong Ma, and Wei Xue. 2021. "A Colon-Targeted Oral Probiotics Delivery System Using an Enzyme-Triggered Fuse-Like Microcapsule." *Advanced Healthcare Materials* 10 (8): 2001953. https://doi.org/10.1002/adhm.202001953.

Coghetto, Chaline Caren, Graziela Brusch Brinques, Nataly Machado Siqueira, Jéssica Pletsch, Rosane Michele Duarte Soares, and Marco Antônio Záchia Ayub. 2016. "Electrospraying Microencapsulation of Lactobacillus Plantarum Enhances Cell Viability under Refrigeration Storage and Simulated Gastric and Intestinal Fluids." *Journal of Functional Foods* 24: 316–26. https://doi.org/10.1016/j.jff.2016.03.036.

Cook, Michael T, Teedah Saratoon, George Tzortzis, Alexander Edwards, Dimitris Charalampopoulos, and Vitaliy V Khutoryanskiy. 2013. "CLSM Method for the Dynamic Observation of PH Change within Polymer Matrices for Oral Delivery." *Biomacromolecules* 14 (2): 387–93. https://doi.org/10.1021/bm301569r.

Dafe, Alireza, Hossein Etemadi, Azita Dilmaghani, and Gholam Reza Mahdavinia. 2017. "Investigation of Pectin/Starch Hydrogel as a Carrier for Oral Delivery of Probiotic Bacteria." *International Journal of Biological Macromolecules* 97 (April): 536–43. https://doi.org/10.1016/J.IJBIOMAC.2017.01.060

Decher, G, J D Hong, and J Schmitt. 1992. "Buildup of Ultrathin Multilayer Films by a Self-Assembly Process: III. Consecutively Alternating Adsorption of Anionic and Cationic Polyelectrolytes on Charged Surfaces." *Thin Solid Films* 210–211: 831–35. https://doi.org/10.1016/0040-6090(92)90417-A.

Depypere, F, P Van Oostveldt, J G Pieters, and K Dewettinck. 2009. "Quantification of Microparticle Coating Quality by Confocal Laser Scanning Microscopy (CLSM)." *European Journal of Pharmaceutics and Biopharmaceutics* 73 (1): 179–86. https://doi.org/10.1016/j.ejpb.2009.04.007.

Dewettinck, K, and A Huyghebaert. 1999. "Fluidized Bed Coating in Food Technology." *Trends in Food Science & Technology* 10 (4): 163–68. https://doi.org/10.1016/S0924-2244(99)00041-2.

Etchepare, Mariana de Araújo, Juliano Smanioto Barin, Alexandre José Cichoski, Eduardo Jacob-Lopes, Roger Wagner, Leadir Lucy Martins Fries, and Cristiano Ragagnin de Menezes. 2015. "Microencapsulation of Probiotics Using Sodium Alginate." *Ciência Rural*. scielo.

Ghorani, Behrouz, and Nick Tucker. 2015. "Fundamentals of Electrospinning as a Novel Delivery Vehicle for Bioactive Compounds in Food Nanotechnology." *Food Hydrocolloids* 51: 227–40. https://doi.org/10.1016/j.foodhyd.2015.05.024.

González-Ferrero, Carolina, Juan Manuel Irache, Beatriz Marín-Calvo, Leticia Ortiz-Romero, Raquel Virto-Resano, and Carlos Javier González-Navarro. 2020. "Encapsulation of Probiotics in Soybean Protein-Based Microparticles Preserves Viable Cell Concentration in Foods All along the Production and Storage Processes." *Journal of Microencapsulation* 37 (3): 242–53. https://doi.org/10.1080/02652048.2020.1724203.

Hills, B P, J Godward, M Debatty, L Barras, C P Saturio, and C Ouwerx. 2000. "NMR Studies of Calcium Induced Alginate Gelation. Part II. The Internal Bead Structure." *Magnetic Resonance in Chemistry* 38 (9): 719–28. https://doi.org/10.1002/1097-458X(200009)38:9<719::AID-MRC739>3.0.CO;2-M.

Iqbal, Rabia, Atif Liaqat, Iqra Yasmin, Samreen Ahsan, Muhmmad Farhan Jahangir Chughtai, Saira Tanweer, Tariq Mehmood, Saima Tehseen, Wahab Ali Khan, Muhammad Nadeem, Assam Bin Tahir, Adnan Khaliq. 2022. "Double Layered Encapsulation to Immobilize *Bifidobacterium Bifidum* ATCC 35914 in Polysaccharide-Protein Matrices and Their Viability in Set Type Yoghurt." *Journal of Food Processing and Preservation* 46 (8): e16748. https://doi.org/10.1111/jfpp.16748.

Jaworek, A, A T Sobczyk, and A Krupa. 2018. "Electrospray Application to Powder Production and Surface Coating." *Journal of Aerosol Science* 125: 57–92. https://doi.org/10.1016/j.jaerosci.2018.04.006.

Jeon, Hyeon Ji, Seung-hwan You, Mijin Kwon, Minhye Shin, Sang-Kyu Kim, and Young Hoon Jung. 2023. "Multilayer Coatings Containing Red Ginseng Dietary Fibre Improve the Survivability and Stability of Probiotic Bacteria." *International Journal of Food Science & Technology* 58 (3): 1497–1505.

Ji, Rui, Jiahui Wu, Junliang Zhang, Tao Wang, Xudong Zhang, Lei Shao, Daijie Chen, and Jian Wang. 2019. "Extending Viability of Bifidobacterium Longum in Chitosan-Coated Alginate Microcapsules Using Emulsification and Internal Gelation Encapsulation Technology." *Frontiers in Microbiology* 10: 1389. https://doi.org/10.3389/fmicb.2019.01389.

Kerdsup, Paramaporn, Passarin Hattayapichat, Juan L Silva, and Sumate Tantratian. 2022. "Survival of Potential Probiotic Isolated from Fermented Tea Leaf and Encapsulated in Multilayer Beads Stored in Makiang (Cleistocalyx Nervosum Var. Paniala) Juice." *Food Bioscience* 50: 102015. https://doi.org/10.1016/j.fbio.2022.102015

Khosravi Zanjani, Mohammad Ali, Babak Ghiassi Tarzi, Anousheh Sharifan, and Nima Mohammadi. 2014. "Microencapsulation of Probiotics by Calcium Alginate-

Gelatinized Starch with Chitosan Coating and Evaluation of Survival in Simulated Human Gastro-Intestinal Condition." *Iranian Journal of Pharmaceutical Research : IJPR* 13 (3): 843–52.

Koh, Wee Y, Xiao X Lim, Thuan-Chew Tan, Rovina Kobun, and Babak Rasti. 2022. "Encapsulated Probiotics: Potential Techniques and Coating Materials for Non-Dairy Food Applications." *Applied Sciences*. https://doi.org/10.3390/app121910005.

Laelorspoen, Nalin, Saowakon Wongsasulak, Tipaporn Yoovidhya, and Sakamon Devahastin. 2014. "Microencapsulation of Lactobacillus Acidophilus in Zein–AlginateCore–Shell Microcapsules via Electrospraying." *Journal of Functional Foods* 7: 342–49. https://doi.org/https:/doi.org/10.1016/j.jff.2014.01.026

Li, Qian, Hongyi Lin, Jing Li, Lu Liu, Jialu Huang, Yi Cao, Tiantian Zhao, David Julian McClements, Jun Chen, Chengmei Liu, Jiyan Liu, Peiyi Shen, Mengzhou Zhou. 2023. "Improving Probiotic (*Lactobacillus Casei*) Viability by Encapsulation in Alginate-Based Microgels: Impact of Polymeric and Colloidal Fillers." *Food Hydrocolloids* 134: 108028. https://doi.org/10.1016/j.foodhyd.2022.108028.

Li, Siyuan, Linlin Fan, Shuangjian Li, Xiaochen Sun, Qingru Di, Hui Zhang, Bin Li, and Xiaoli Liu. 2023. "Validation of Layer-By-Layer Coating as a Procedure to Enhance Lactobacillus Plantarum Survival during In Vitro Digestion, Storage, and Fermentation." *Journal of Agricultural and Food Chemistry* 71 (3): 1701–12. https://doi.org/10.1021/acs.jafc.2c07139

Li, Wenyan, Xuejiao Lei, Hua Feng, Bingyun Li, Jiming Kong, and Malcolm Xing. 2022. "Layer-by-Layer Cell Encapsulation for Drug Delivery: The History, Technique Basis, and Applications." *Pharmaceutics*. https://doi.org/10.3390/pharmaceutics14020297.

Librán, C M, S Castro, and J M Lagaron. 2017. "Encapsulation by Electrospray Coating Atomization of Probiotic Strains." *Innovative Food Science & Emerging Technologies* 39: 216–22.https:/doi.org/10.1016/j.ifset.2016.12.013.

Lvov, Yuri, Gero Decher, and Helmuth Moehwald. 1993. "Assembly, Structural Characterization, and Thermal Behavior of Layer-by-Layer Deposited Ultrathin Films of Poly(Vinyl Sulfate) and Poly(Allylamine)." *Langmuir* 9 (2): 481–86. https://doi.org/10.1021/la00026a020

Mabrouk, Ahmed M, Heba H Salama, Hoda S El Sayed, and Samah M El Sayed. 2021. "Preparation of Symbiotic Whey Protein Gel as a Carrier of Free and Encapsulated Probiotic Bacteria." *Journal of Food Processing and Preservation* 45 (7): e15612. https:/doi.org/10.1111/jfpp.15612

Masoomi Dezfooli, Seyedehsara, Chloe Bonnot, Noemi Gutierrez-Maddox, Andrea C Alfaro, and Ali Seyfoddin. 2022. "Chitosan Coated Alginate Beads as Probiotic Delivery System for New Zealand Black Footed Abalone (Haliotis Iris)." *Journal of Applied Polymer Science* 139 (29): e52626.

Mendes, Ana C, and Ioannis S Chronakis. 2021. "Electrohydrodynamic Encapsulation of Probiotics: A Review." *Food Hydrocolloids* 117: 106688. https:/doi.org/10.1016/j.foodhyd.2021.106688.

Menezes, Maria Fernanda da Silveira Cáceres de, Thaiane Marques da Silva, Mariana de Araújo Etchepare, Bruna de Souza Fonseca, Vandré Pinto Sonza, Cristiane Franco Codevilla, Juliano Smanioto Barin, Cristiane de Bona da Silva, and Cristiano

Ragagnin de Menezes. 2019. "Improvement of the Viability of Probiotics (*Lactobacillus Acidophilus*) by Multilayer Encapsulation." *Ciência Rural*. scielo .

Mirzamani, S S, A R Bassiri, H Tavakolipour, M H Azizi, and M Kargozari. 2021. "Survival of Fluidized Bed Encapsulated Lactobacillus Acidophilus under Simulated Gastro-Intestinal Conditions and Heat Treatment during Bread Baking." *Journal of Food Measurement and Characterization* 15 (6): 5477–84. https://doi.org/10.1007/s11694-021-01108-0.

Moayyedi, Mahsa, Mohammad Hadi Eskandari, Amir Hossein Elhami Rad, Esmaeil Ziaee, Mohammad Hossein Haddad Khodaparast, and Mohammad-Taghi Golmakani. 2018. "Effect of Drying Methods (Electrospraying, Freeze Drying, and Spray Drying) on Survival and Viability of Microencapsulated Lactobacillus Rhamnosus ATCC 7469." *Journal of Functional Foods* 40: 391–99. /https://doi.org/10.1016/j.jff.2017.11.016.

Mogna, Giovanni, and Luca Mogna. 2014. Multilayer microencapsulated probiotic bacteria. US 2014/0370107 A1, issued 2014. https://patents.google.com/patent/WO2013114185A1/en.

Mokarram, R R, S A Mortazavi, M B Habibi Najafi, and F Shahidi. 2009. "The Influence of Multi-Stage Alginate Coating on Survivability of Potential Probiotic Bacteria in Simulated Gastric and Intestinal Juice." *Food Research International* 42 (8): 1040–45. https://doi.org/10.1016/j.foodres.2009.04.023.

Niculescu, Adelina-Gabriela, and Alexandru Mihai Grumezescu. 2022. "Applications of Chitosan-Alginate-Based Nanoparticles-An Up-to-Date Review." *Nanomaterials (Basel, Switzerland)* 12 (2). https://doi.org/10.3390/nano12020186

Pech-Canul, Angel D, David Ortega, Antonio García-Triana, Napoleón González-Silva, and Rosa L Solis-Oviedo. 2020. "A Brief Review of Edible Coating Materials for the Microencapsulation of Probiotics." *Coatings*. https://doi.org/10.3390/coatings10030197.

Pedroso, D L, M Dogenski, M Thomazini, R J B Heinemann, and C S Favaro-Trindade. 2013. "Microencapsulation of Bifidobacterium Animalis Subsp. Lactis and Lactobacillus Acidophilus in Cocoa Butter Using Spray Chilling Technology." *Brazilian Journal of Microbiology*. scielo.

Phuong Ta, Linh, Erika Bujna, Szilárd Kun, Dimitris Charalampopoulos, and Vitaliy V Khutoryanskiy. 2021. "Electrosprayed Mucoadhesive Alginate-Chitosan Microcapsules for Gastrointestinal Delivery of Probiotics." *International Journal of Pharmaceutics* 597: 120342. https://doi.org/10.1016/j.ijpharm.2021.120342.

Pitigraisorn, Panithan, Kantaphong Srichaisupakit, Nutnicha Wongpadungkiat, and Saowakon Wongsasulak. 2017. "Encapsulation of Lactobacillus Acidophilus in Moist-Heat-Resistant Multilayered Microcapsules." *Journal of Food Engineering* 192: 11–18. https://doi.org/10.1016/j.jfoodeng.2016.07.022.

Poddar, Devastotra, Jon Palmer, Shantanu Das, Manju Gaare, Arup Nag, and Harjinder Singh. 2022. "Effect of Fluidized Bed Drying, Matrix Constituents, and Structure on the Viability of Probiotic Lactobacillus Paracasei ATCC 55544 during Storage at 4 ° C, 25 ° C, and 37 ° C." *Microorganisms*. https://doi.org/10.3390/microorganisms10010074.

Pupa, Pawiya, Prasert Apiwatsiri, Wandee Sirichokchatchawan, Nopadon Pirarat, Nongnuj Muangsin, Asad Ali Shah, and Nuvee Prapasarakul. 2021. "The Efficacy of Three

Double-Microencapsulation Methods for Preservation of Probiotic Bacteria." *Scientific Reports* 11 (1): 13753. https://doi.org/10.1038/s41598-021-93263-z.

Qi, Xiaoxi, Senay Simsek, Jae-Bom Ohm, Bingcan Chen, and Jiajia Rao. 2020. "Viability of Lactobacillus Rhamnosus GG Microencapsulated in Alginate/Chitosan Hydrogel Particles during Storage and Simulated Gastrointestinal Digestion: Role of Chitosan Molecular Weight." *Soft Matter* 16 (7): 1877–87. https://doi.org/10.1039/C9SM02387A.

Santos, Matheus A S, and Mariana T C Machado. 2021. "Coated Alginate–Chitosan Particles to Improve the Stability of Probiotic Yeast." *International Journal of Food Science & Technology* 56 (5): 2122–31. https://doi.org/10.1111/ijfs.14829.

Sathyabama, S, M Ranjith Kumar, P Bruntha devi, R Vijayabharathi, and V Brindha priyadharisini. 2014. "Co-Encapsulation of Probiotics with Prebiotics on Alginate Matrix and Its Effect on Viability in Simulated Gastric Environment." *LWT - Food Science and Technology* 57 (1): 419–25. https://doi.org/10.1016/j.lwt.2013.12.024.

Schär-Zammaretti, Prisca, and Job Ubbink. 2003. "The Cell Wall of Lactic Acid Bacteria: Surface Constituents and Macromolecular Conformations." *Biophysical Journal* 85 (6): 4076–92. https://doi.org/10.1016/S0006-3495(03)74820-6

Schell, Daniel, and Christopher Beermann. 2014. "Fluidized Bed Microencapsulation of Lactobacillus Reuteri with Sweet Whey and Shellac for Improved Acid Resistance and In-Vitro Gastro-Intestinal Survival." *Food Research International* 62: 308–14. https://doi.org/10.1016/j.foodres.2014.03.016.

Silva, Karen Cristina Guedes, Ana Isabel Bourbon, Lorenzo Pastrana, and Ana Carla Kawazoe Sato. 2020. "Emulsion-Filled Hydrogels for Food Applications: Influence of PH on Emulsion Stability and a Coating on Microgel Protection." *Food & Function.* https://doi.org/10.1039/D0FO01198C

Talebian, Sepehr, Timothy Schofield, Peter Valtchev, Aaron Schindeler, John M Kavanagh, Qayyum Adil, and Fariba Dehghani. 2022. "Biopolymer-Based Multilayer Microparticles for Probiotic Delivery to Colon." *Advanced Healthcare Materials* 11 (11): 2102487.https:/doi.org/10.1002/adhm.202102487.

Thomas, Midhun Ben, Mahalakshmi Vaidyanathan, Krishna Radhakrishnan, and Ashok M Raichur. 2014. "Enhanced Viability of Probiotic Saccharomyces Boulardii Encapsulated by Layer-by-Layer Approach in PH Responsive Chitosan–Dextran Sulfate Polyelectrolytes." *Journal of Food Engineering* 136: 1–8. https://doi.org/10.1016/j.jfoodeng.2014.03.015.

Yeung, Timothy W, Elif F Üçok, Kendra A Tiani, David J McClements, and David A Sela. 2016. "Microencapsulation in Alginate and Chitosan Microgels to Enhance Viability of Bifidobacterium Longum for Oral Delivery." *Frontiers in Microbiology* 7: 494. https://doi.org/10.3389/fmicb.2016.00494.

Yucel Falco, Cigdem, Federico Amadei, Surender K Dhayal, Marité Cárdenas, Motomu Tanaka, and Jens Risbo. 2019. "Hybrid Coating of Alginate Microbeads Based on Protein-Biopolymer Multilayers for Encapsulation of Probiotics." *Biotechnology Progress* 35 (3): e2806. https:/doi.org/10.1002/btpr.2806

Zaeim, Davood, Mahboobe Sarabi-Jamab, Behrouz Ghorani, and Rassoul Kadkhodaee. 2019. "Double Layer Co-Encapsulation of Probiotics and Prebiotics by Electro-

Hydrodynamic Atomization." *LWT* 110: 102–9. https://doi.org/10.1016/j.lwt.2019.04.040

Zaeim, Davood, Mahboobe Sarabi-Jamab, Behrouz Ghorani, Rassoul Kadkhodaee, and R Hans Tromp. 2017. "Electrospray Assisted Fabrication of Hydrogel Microcapsules by Single- and Double-Stage Procedures for Encapsulation of Probiotics." *Food and Bioproducts Processing* 102: 250–59. https://doi.org/10.1016/j.fbp.2017.01.004.

Zaghari, Leila, Alireza Basiri, and Somayeh Rahimi. 2020. "Preparation and Characterization of Double-Coated Probiotic Bacteria via a Fluid-Bed Process: A Case Study on Lactobacillus Reuteri," International Journal of Food Engineering, 16 (9). https://doi.org/doi:10.1515/ijfe-2019-0384

Zhao, Shuang, Frank Caruso, Lars Dähne, Gero Decher, Bruno G De Geest, Jinchen Fan, Neus Feliu, Yury Gogotsi, Paula T. Hammond, Mark C. Hersam, Ali Khademhosseini, Nicholas Kotov, Stefano Leporatti, Yan Li, Fred Lisdat, Luis M. Liz-Marzán, Sergio Moya, Paul Mulvaney, Andrey L. Rogach, Sathi Roy, Dmitry G. Shchukin, Andre G. Skirtach, Molly M. Stevens, Gleb B. Sukhorukov, Paul S. Weiss, Zhao Yue, Dingcheng Zhu, and Wolfgang J. Parak. 2019. "The Future of Layer-by-Layer Assembly: A Tribute to ACS Nano Associate Editor Helmuth Möhwald." *ACS Nano* 13 (6): 6151–69. https://doi.org/10.1021/acsnano.9b03326

## Chapter 9

# *In Vitro* and *In Vivo* Digestibility of Microparticles Containing Probiotics

**Maria José do Amaral e Paiva*** 
**Nataly de Almeida Costa**
**and Vanessa Caroline de Oliveira**
Department of Food Technology (DTA), Federal University of Viçosa (UFV), Viçosa, Minas Gerais, Brazil

### Abstract

To promote health benefits, probiotics need to survive adverse conditions in the carrier matrix and during passage through the gastrointestinal system. Thus, the choice of probiotic, as well as the microencapsulation technique and the wall materials used need to protect these cells when exposed to variations in pH, stomach acids, and the presence of different enzymes until reaching the site of action in sufficient quantities ($>10^6$) to promote the desired benefits. This chapter describes the entire digestive process and the influence of each step on the survival of probiotics. *In vitro* and in vivo digestion as well as the different models used to evaluate the efficiency of microencapsulation techniques and the survival of probiotics were described. Studies using free or microencapsulated probiotics have evaluated the survival of these microorganisms during passage through the gastrointestinal system. Finally, this chapter brings together one of the main points of the application of probiotics in food matrices for delivery at the appropriate site of action and in sufficient quantities to promote the intended benefits.

* Corresponding Author's Email: maria.j.amaral@ufv.br.

In: Microencapsulation of Probiotics
Editors: D. de Almeida Paula, K. Cristina Guedes Silva and N. de Almeida Costa
ISBN: 979-8-88697-979-4

**Keywords:** gastrointestinal system, digestive enzymes, probiotics, viability

## Introduction

The *in vitro* and *in vivo* digestibility of microparticles containing probiotic microorganisms is an important factor in assessing the stability and protection provided by the chosen microencapsulation technique and, consequently, the wall materials used. In addition to evaluating the protection and stability of the probiotic, it is a way to evaluate and improve the efficiency of microcapsules and promote a controlled release of these microorganisms in the gastrointestinal tract in sufficient quantities to promote health benefits.

Probiotic microorganisms are sensitive to the adverse conditions found in the gastrointestinal tract, such as the presence of strongly acidic gastric fluids, digestive enzymes, and bile salts in the intestinal phases, which can cause a reduction in the survival of probiotic microorganisms, resulting in limited colonization and low multiplication of probiotics in the intestine.

Thus, *in vitro* and in vivo studies are used to carry out the simulation and meet the need to know aspects related to the route that the food follows after ingestion and how the microencapsulated probiotics interact with the constituents present in that food. Thus, this chapter aims to describe the digestive process and the survival of the probiotic during this process, to detail the behavior of microcapsules in vivo and *in vitro* studies and the different models used.

## Digestive System and Probiotic Survival

The digestive system is formed by the digestive tract and the organs called accessories, its morphology is represented by a tube made up of muscles, it is approximately 9 m long. The digestive tract consists of the mouth, pharynx, esophagus, stomach, small intestine, and large intestine and the accessory organs are the teeth, tongue, salivary glands, liver, gallbladder, and pancreas. The system is open at both ends, the stomach and intestines make up the gastrointestinal tract (Saladin 2016).

The digestive process can be divided into two main physical-chemical segments that happen at the same time, one of them is the disintegration of the food reducing it into smaller particles and the other is changes promoted by the enzymes that hydrolyze the macromolecules in sizes smaller enough to be absorbed by epithelial cells (Guerra et al. 2012).

The human digestive system is highly complex, from the moment food is ingested, it begins to undergo transformation and metabolization processes. It is through digestion that food provides the body with the nutrients necessary for human survival (Dupont et al. 2019).

In this multi-compartmental system, throughout the digestion process, there is a constant interaction between each compartment with its specific attributions and the food. Concomitantly to the mechanical action, there is the release and action of enzymes and bile salts so that the hydrolysis of nutrients occurs and at the end of this physiological process, the nutrients have been released and only a residue remains to be discarded (LI et al., 2023).

In the mouth, when food is ingested, it undergoes a mechanical effect and is fragmented into smaller particles under constant saliva soaking, this is where some enzymes present in saliva already start to act. The finalization of food disintegration processes takes place in the stomach through peristalsis movements. In the stomach, there is a drastic reduction in pH and a strong action of enzymes, this is where particles of different sizes are segregated, with only those smaller than 1 mm remaining (Guerra et al. 2012). The separation of particle sizes takes place by the mechanism of recirculation and differential emptying. In the small intestine, several enzymatic activities occur, including fragmentation of macromolecules and movements of contraction and distension of the mucosal villi that are most responsible for absorption. In the colon, food that has not been digested is fermented and water is absorbed. There are strong and constant changes in pH and enzyme activities during the process (Figure 1) (Guerra et al. 2012; Marze, 2017).

In addition to enzymes, pH changes, movements, and fluids, the digestive system has a powerful partnership to assist in digestion, vitamin synthesis, and release of bioactive compounds, among others, which are the microorganisms or microbiota present, especially those that colonize the intestinal mucosa. In the intestine, there are about 100 trillion microorganisms, in the many folds of the tissue of the colon approximately four thousand different species are housed (Collen 2016).

When dysbiosis occurs, that is, the imbalance of the intestinal microbiota, the person suffers from several health problems, among them metabolic and developmental disorders, inflammation, food intolerances, irritable bowel syndrome, and even psychiatric illnesses (Langdon, Crook, and Dantas 2016).

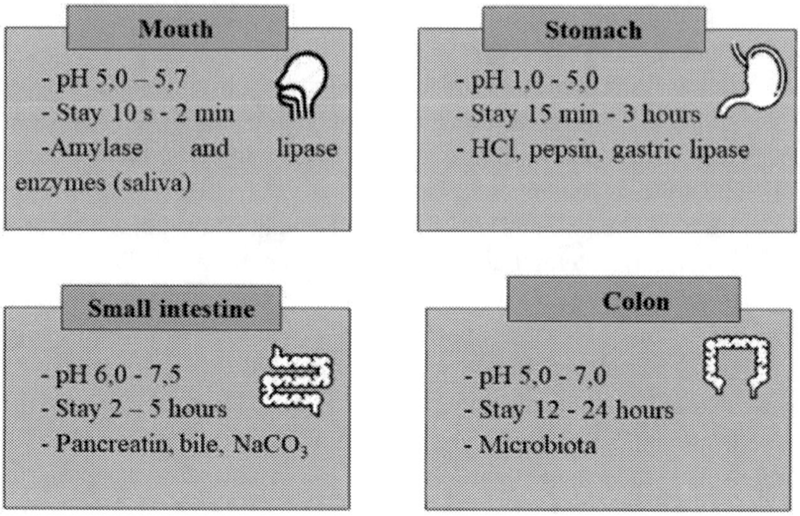


**Figure 1**. Some physicochemical and biological characteristics of the digestive system.

One of the ways to stimulate the balance of the intestinal microbiota is the addition of probiotic microorganisms to foods to modulate this microbiota. Well-founded studies over the years show that foods are excellent carriers (Matejceková et al. 2017; Adebayo-tayo, Olomitutu, Adebami, 2021; Rasika et al. 2021) and also show that probiotic microorganisms promote several health benefits (Bellis, Sisto, and Lavermicocca 2021). More recently, with the deepening of studies on the relationship between the intestine and the brain forming an axis, the effect of probiotics on emotional psychiatric illnesses has been proven (Bagga et al. 2018; Terpou et al. 2019).

Baião et al. (2022), in a double-blind study, multispecies probiotic microorganisms were administered to people with mild to moderate depression who were not taking medication and found that there was an improvement in depression compared to those who took a placebo. The term psychobiotic came about due to studies showing the effect of probiotic bacteria on psychiatric illnesses.

The use of probiotics for four weeks in healthy humans, with a similar body mass index and the same average age, was related to a positive change in behavior and memory patterns, showing changes in brain activation patterns influenced by the gut-brain axis, measured through emotions. The alterations happened at the same time as the small changes that occurred in the intestinal microbiota, evaluated before and after the administration of the probiotic bacteria (Bagga et al. 2018).

But even with all the benefits provided by probiotics, their addition to food matrices and delivery to the site of action in the human intestine poses several challenges to the viability and survival of the strains. The challenges already begin during processing and storage through intrinsic factors (pH, titratable acidity, presence of oxygen, sugar salt, hydrogen peroxide, substances with bactericidal effect, and additives such as flavorings and colorings) and extrinsic factors such as thermal treatment, temperature storage conditions and compounds present in packaging materials (Thorakkattu et al. 2022). However, the challenges do not stop there, there are also those related to the digestion process.

In the mouth, probiotic microorganisms face the microbicidal action of saliva and exposure to oxygen. During the entire transit through the gastrointestinal tract, there is exposure to different acidity conditions, the effect of bile salts, pancreatic secretion, and even adherence to the present mucus layer (Terpou et al. 2019). This hostile environment of the gastrointestinal tract makes probiotic survival difficult and often prevents cells from arriving viable and with the appropriate count at the specific location to promote benefits to the host (Mao et al. 2019), thus ensuring that the minimum amount necessary is delivered, of $10^6$ CFU per daily serving of food is a challenge.

Therefore, one of the main criteria for choosing or selecting probiotics is their ability to survive in acidic conditions precisely because of the low pH found during transit through the gastrointestinal system. This is because, in acidic environments, damage to the cell membrane and DNA of the microorganism can occur. (Lillo-Pérez et al. 2021).

Different food matrices are used as carriers of probiotics, each matrix has unique characteristics and has a different influence on the viability and survival of microorganisms during the processing, storage, and digestion stages. Some complementary mechanisms are applied to protect probiotic bacteria, such as the combination of two or more food matrices, the addition of prebiotics, and the incorporation of probiotic cells in hydrocolloids (Pavli et al. 2017; Ningtyas et al. 2019).

And there are also survival mechanisms developed by the microorganisms themselves in response to the stressful environment, which can be induced the emergence of more tolerant strains. Activation of these mechanisms can occur through the cultivation of strains under conditions of sublethal stress. Some types of stress are temperature changes, the creation of acidic environments, osmotic pressure, high pressure, the presence of oxygen, and even the absence of nutrients. (Terpou et al. 2019).

But the most studied mechanism lately to increase the survival of probiotic bacteria during processing, storage period, and also during passage through the gastrointestinal tract is microencapsulation. This technology is well-defined as being of great importance for keeping cells viable. Different microencapsulation techniques and various materials can be used to obtain microcapsules (De Prisco; Mauriello, 2016; Mao et al. 2019). *In vitro* and *in vivo* tests are carried out to show the effectiveness of microcapsules in protecting probiotic microorganisms, allowing their controlled release at the ideal site of action in sufficient quantities to exert the expected effect on human health.

## Digestion *In Vitro* and *In Vivo*

*In vitro,* studies of the digestion process were developed due to the need to know aspects related to the route that food follows from ingestion and how the body interacts with the constituents present in that food. Among the main reasons for this are the growing number of people with food-related illnesses; the need to prepare new foods to serve specific groups of people, adding to identifying the bioaccessibility and bioavailability of compounds, nutrients, and vitamins present in food, the kinetics of particle disintegration and evaluation of the survival of probiotic microorganisms (Brodkorb et al. 2019; Fioravanti, Morgano, 2021).

Despite the high complexity of the digestion process, the simplest way to understand it is through models with simulation, since studies with *in vivo* models, whether human or animal, are costly, often requiring the use of cannulas or probes placed in the organs, digestive devices to access contents of the gastrointestinal tract, which, however, are invasive and for which several ethical limitations are imposed (Hur et al. 2011; Wickham et al. 2012; Dupont et al. 2019).

In addition to being less expensive and without ethical restrictions compared to *in vivo* models, *in vitro* simulation offers the possibility of greater reproducibility by being subjected to more easily controlled conditions, in addition to allowing the analysis of a greater number of samples of interest in a short period of time (Minekus et al. 2014). However, even though several simulation models have already been created, they cannot fully reproduce all the natural processes of human digestion, studies related to bioavailability, bioaccessibility, emptying efficiency, changes in structural characteristics, and intestinal fluids are still necessary (Marze, 2017; Brodkorb et al. 2019), even so, they are the ones that best represent the digestive system for studies.

Methods applying *in vitro* gastrointestinal simulation are used for several purposes and one of them is to measure the efficiency of microcapsules in protecting probiotic microorganisms. Due to the growing search for functional foods, studies with the addition of probiotics have increased, especially in plant matrices. One of the ways to confirm whether the microencapsulated cells will remain viable is by subjecting them to the conditions they will face in the gastrointestinal tract.

Costa et al. (2023), researching the protective effect of microcapsules on *L. plantarum,* used maltodextrin, sweet potato flour, pectin, and gelatin as wall material and obtained counts above 8 Log CFU.g$^{-1}$ after the *in vitro* gastrointestinal mock test. When heated, sweet potato starch is converted into resistant starch that is not digested in the human digestive system, offering protection and serving as a fermentation substrate for microorganisms when released into the colon.

In a study comparing the viability and survival of microencapsulated and free *Lactiplantibacillus plantarum* cells in mixed carrot and acerola juice stored for 28 days at 5°C, it was verified that there was no alteration in the physical-chemical and functional characteristics of the juice that contained the cells of the microencapsulated microorganism. Despite the viability during storage being the same for both treatments, when *in vitro* gastrointestinal simulation, viability, and survival were much higher in juice with microcapsules compared to free cells, with 7.20 Log CFU.mL$^{-1}$ and 4.07 Log CFU.mL$^{-1}$ respectively (Paula et al. 2021). In addition to protecting the microorganism, the microcapsules also prevent it from promoting changes in the physicochemical attributes of the product caused by fermentation.

*L. salivarius* spp. *salivarius* microencapsulated by High-Pressure Homogenization (HPH) at 70 MPa in sodium alginate, Tween 80, and sunflower oil added to tangerine juice was compared to juice containing cells free of microorganisms. Evaluations were carried out during 1, 3, 7, and 10 days of storage, the gastrointestinal survival analyses showed greater survival of microencapsulated microorganisms, respectively 9.09; 7.93; 6;87; 6.64 Log CFU.mL$^{-1}$ for microencapsulated and 8.08; 7.53; 6.14 and 5.65 Log CFU.mL$^{-1}$ for free cells. Microencapsulated cells had an approximate reduction of 50% while free cells were 75-85% during the gastrointestinal simulation (Calabuig-Jiménez et al. 2019).

Following the study, tangerine juice containing *L. salivarius* spp. *salivarius* free and microencapsulated was used to impregnate apple slices that were then dried and evaluated during the 30-day shelf life. Microencapsulated samples showed less pH variation, as encapsulation reduces the fermentative activity of microorganisms. Up to 14 days of storage there was no difference between treatments, after this period the encapsulated microorganisms suffered less degradation effect, also showing greater resistance to gastrointestinal simulation (Betoret et al. 2019).

The survival of *Lactobacillus casei* and *Lactobacillus acidophilus* was evaluated after addition to *Antidesma bunius L.* juice containing maltodextrin, *Tiliacora triandra* gum, and/or inulin for microencapsulation. The comparisons were in the powdered probiotic juice containing or not *Tiliacora triandra* gum, inulin, and juice with free cells. In the samples where the microorganisms were microencapsulated by maltodextrin and *Tiliacora triandra* gum, survival was higher during spray drying, this treatment also obtained greater survival in the *in vitro* gastrointestinal simulation, presenting a higher count for *Lactobacillus acidophilus* (Chaikham, Kemsawasd, Seesuriyachan, 2017).

The microencapsulation of *L. rhamnosus* by agave fructans, gum arabic, maltodextrin, trehalose, and inulin promoted encapsulation efficiency of around 90%, with preservation of the structure of the microcapsules. The combination that promoted the best preservation of microorganisms was gum arabic/trehalose after 8 weeks of storage at 4°C, with a reduction of only 1 log cycle. This combination was also the one that best protected the probiotics during *in vitro* gastrointestinal simulation. Free *L. rhamnosus* cells had an average reduction of 5.25 Log $CFU.g^{-1}$ in the presence of gastric juice, whereas for microencapsulated cells the average reduction was 2.65-4.50 Log $CFU.g^{-1}$ (Barajas-Álvarez, González-Ávila, and Espinosa-Andrews 2022).

Through digestion simulation methods, it is possible to verify whether the wall materials used to microencapsulate the probiotics are efficient in offering the necessary protection to the microorganism so that it is released in a controlled manner where it will exert the expected benefits to the host (De Prisco and Mauriello 2016).

There are models of different *in vitro* methods that simulate the stages of the process by which the food is submitted to the digestive system. These models are classified as static and dynamic mono or multicompartment. In the static model, the movement takes place in an orbital manner, while in the dynamic model, much closer to what happens in human digestion, the movement of the flow is more complex. Simulation models are proven screening tools, and the most widely used among the described methods is the static one (Dupont et al. 2019).

## *In Vitro* Static Digestion Model

Static digestive simulation models present simpler protocols to be applied and are the most described in the literature. The enzyme concentration and pH are kept constant according to each step. Samples are placed in bioreactors and the medium is altered according to physicochemical characteristics and predetermined enzymatic activities for each phase. It is a model suitable only for evaluating the result of the digestion stages, as it does not work with the kinetics of the process. The advantages and disadvantages of the model are described in Table 1 (Brodkorb et al. 2019; Dupont et al. 2019).

Several static digestion protocols were created, but because there were many discrepancies between the models, a group of more than 45 researchers from INFOGEST (an international network that studies the interaction of food with the body through the digestive system) carried out a harmonization of the experimental conditions that simulate the digestion of adults. For validation and confirmation of harmonization, tests were developed in different laboratories in several countries using skimmed milk powder as a standardization matrix, and the degradation of proteins, generation of peptides, and free amino acids were analyzed. For *in vivo* investigation, analyses were carried out in eight pigs, which were fed three times with skimmed milk powder 6, 3, and 1.5 hours before sacrifice. Samples were collected from the stomach, duodenum, small intestine, and large intestine to evaluate the same variables analyzed *in vitro*. The results showed that there was an experimental improvement in the comparison between laboratories and that the gastric and intestinal stages of the simulation represent the *in vivo* reality in pigs about protein digestion (Bohn et al. 2018).

The static model can be used both to simulate all phases in total digestion or only for a certain phase of interest, as an example, the protocol described by Bedani, Rossi, Saad (2013) can be mentioned, where food aliquots, in liquid form, are submitted directly to the gastric phase, proceeding to enteric phase I and enteric phase II.

**Table 1.** Advantages and disadvantages of static digestion simulation

| | Advantages | Disadvantages |
|---|---|---|
| Static model | Simpler execution | Enzymatic assays and standardized pH |
| | Easy playback | Absence of gradual addition of gastric fluid |
| | Easy assessment of the different stages | Absence of gradual emptying of each compartment |
| | Reliability | |
| | Lower cost | |

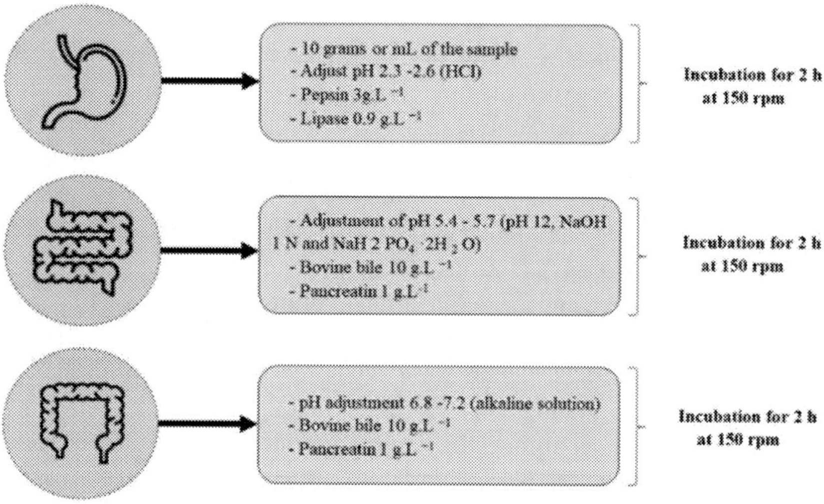


**Figure 2**. Flowchart of the static digestive simulation protocol in three phases only.

In this model of Bedani, Rossi, and Saad (2013), 10 grams or mL of the sample are prepared, and diluted in 90 mL of 0.85% NaCl solution. To simulate the gastric phase, aliquots of the dilution are transferred to sterile flasks, and pepsin and lipase are added at a concentration of 3 $g.L^{-1}$ and 0.9 $mg.L^{-1}$, respectively. The pH adjustment with 1N HCl is carried out to the range between 2.3-2.6. The flasks with the samples are incubated for 2 h at 37℃ under agitation at 150 rpm in a shaker. After this period, a simulation of what happens in the small intestine is performed, the pH is adjusted to 5.4-5.7 with sodium phosphate pH 12 (150 mL of 1N NaOH; 14 g of $NaH_2PO_4$.

$2H_2O$), the bile and pancreatin enzymes in the proportion of 5 $g.L^{-1}$ and 1.6 $g.L^{-1}$, followed by a new incubation for another 2 h at 37°C under agitation at 150 rpm. At the end of this period, what happens in the large intestine is simulated, where the pH is adjusted to 6.8-7.2 using sodium phosphate solution pH 12 containing bile (7.95 $g.L^{-1}$) and pancreatin (0.79 $g.L^{-1}$), another incubate ion period is carried out under the same conditions as before, for a total of 6 hours (Figure 2).

Depending on the objectives proposed with the gastric simulation research, only the final product can be analyzed or samples can be taken at the end of each phase to be submitted for analysis. To interrupt the digestion process, it is necessary to inhibit or neutralize the enzymatic activity. Some alternatives are immediate freezing in liquid nitrogen. To interrupt the gastric phase, sodium bicarbonate at a concentration of 0.5 M can be added before freezing and lyophilization of the samples afterward. Another possibility is the addition of a protease inhibitor (4-(2-aminoethyl) benzene sulfonyl fluoride hydrochloride) and then subjected to freezing and subsequent lyophilization (Minekus et al. 2014).

To assess probiotic viability and survival in the gastrointestinal simulation, which is performed at the end of each phase and the end of the entire process, the neutralization of enzymatic activity can be performed with the addition of buffered saline solution (Würth et al. 2015) before plating for subsequent CFU counting.

Static models simplify the digestive process and make research related to digestion products more accessible, which is why it is the most used and mentioned in the literature (Dupont et al. 2019), but the study of the behavior of foods and their constituents cannot be well understood through these models because they do not use more realistic dimensions of fluid kinetics and physiological changes that occur during the process (Wickham et al., 2012).

## Dynamic *In Vitro* Digestion Model

Even with all the advantages associated with the use of static models of gastrointestinal simulation, they are flawed in evaluating the transformations that occur in the process dynamically, such as pH variations, the gradual addition of enzymes and gastric fluid, and the continuity of emptying of the gastrointestinal tract gastric compartments. Dynamic digestion simulation models are more complete because they mimic the physiological conditions of monogastric. In these protocols, the gradual addition of enzymes occurs in each phase according to their concentrations, and the pH changes the process also gradually (Minekus et al. 2014).

The dynamic models were created based on the need to know the kinetics of fluids, movements, and biochemical changes in the human digestive phases. This model came to meet the demand for analysis of more delicate physiological processes such as bioaccessibility to food constituents, the effect resulting from the structure of foods on bioaccessibility, benefits arising from functional foods, such as the survival of probiotics added to foods (Dupont et al. 2019), some models are written in Table 2.

Among the described methods, the SHIME - Human Intestinal Microbial Ecosystem Simulator is a more complete and complex dynamic system simulating even the human gastrointestinal microbiota of both adults and healthy babies and the elderly (Van den Abbeele et al. 2012).

One of the simplest dynamic models is the Human Gastric Simulator (HGC) composed of the gastric compartment chamber (Figure 3). The mechanical drive that mimics the contraction waves is carried out through belts. The drive shaft is connected to the mixing head, the speed is kept in the range of 2 to 180 rpm. Teflon rollers are attached to the belts and as the belts move the rollers collide with the walls of the latex compartment. With programming to propagate three contractions per minute. A peristaltic pump is responsible for gastric emptying, the digest is sieved through a mesh bag that lines the latex compartment, and only the smallest particles are sieved. The fluids are released at the bottom of the cavity (10 to 15 cm in height from the bottom) via a 0.86 mm Teflon tube with a flow rate of 0.03 to 8.2 mL/min. The system is placed inside a chamber insulated by plastic foam and the temperature control, 37°C, is done through a thermostat coupled to two 60 W lamps (Kong, Singh, 2010; Ferrua, Singh, 2015).

Dynamic *in vitro* models that allow inoculation by microorganisms or the presence of a mucus layer facilitate studies showing the interaction of probiotic microorganisms with the intestinal microbiota, such as the study by Gil-Sánchez et al. (2020), who inoculated *L. plantarum* CLC-17 in a dynamic gastrointestinal simulator with a fresh fecal solution to establish and stabilize the microbiota, substrates were added before implanting the probiotic. The system was fed 3 times a day, and the whole experiment lasted 45 days *L. plantarum* was successfully established at 7.39 Log CFU.mL$^{-1}$ and there was an increase in viability during the period that the supplementation was carried out even under the conditions of gastric pH, presence of bile salts, and other enzymes. However, when the administration was interrupted, the microorganism counts dropped, and at 4 days of washout they were reduced to 5.31 Log CFU.mL$^{-1}$ (Derrien and Hylckama Vlieg 2015).

*In vitro* models are widely applied in research and are fundamental for understanding, in a more simplified way, the real circumstances of digestion (Bellmann et al. 2016). However, despite being proven screening tools, they cannot faithfully reproduce the physical-chemical and biological processes of human digestion with all their peculiarities, so *in vivo* tests are necessary (Luo et al. 2020). Even considering the obstacles to these types of tests, they are the gold standard for validating mimicry methods.

For studies involving probiotic microorganisms, *in vivo,* assays make it possible to more efficiently identify the benefits of probiotics. In addition to studies involving humans, animal models can be used, mainly pigs, rats, and mice. (Barajas-Álvarez, González-Ávila, and Espinosa-Andrews 2021).

Among the *in vivo* assays, microcapsule rupture tests containing probiotic bacteria can be mentioned, microcapsules must protect the microorganism but cannot be insurmountable barriers (Würth et al. 2015), the rupture of microcapsules at the ideal time and place is essential for probiotics to provide the described health benefits (Costa et al. 2023).

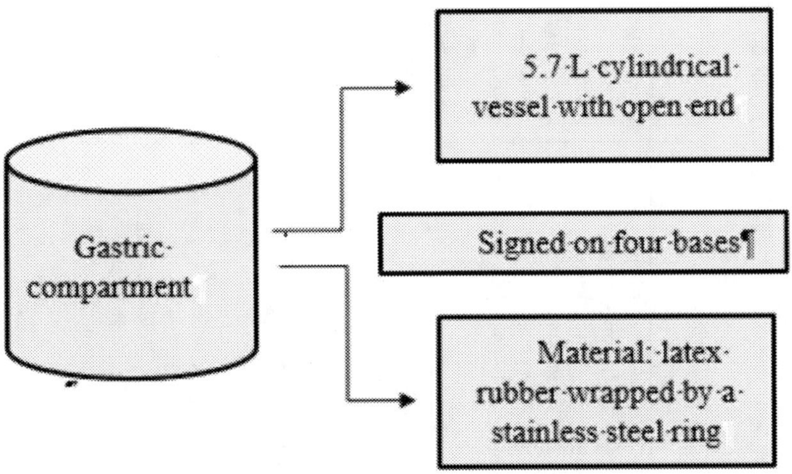


**Figure 3.** Representation of the gastric chamber of the HGC model.

**Table 2.** Dynamic digestion simulation models

| Model | Simulated phase | Particularities | Benefits | Disadvantages | *In vivo* validation | References |
|---|---|---|---|---|---|---|
| SHIME | Mouth, stomach, duodenum, jejunum, ileum, and large intestine | Addition of mucous layer for microbial community development, gastrointestinal simulation of healthy adults, infants, and elderly | Computerized technological model. Possibility to analyze ingredients and products. Presence of the mucosa for interaction studies with probiotics and pathogens | Not accurately mimicking physiological functions | *in vivo* data of bacterial strains used to indicate microbial colonization, the occurrence of enzyme activity, presence, and amount of short-chain fatty acids. Response to pharmaceutical drugs. Interaction with probiotics and metabolization of prebiotics. | (Van den Abbeele et al. 2010;Van den Abbeele et al. 2012; Van den Abbeele et al. 2013) |
| TIM | Stomach, small and large intestine | Compartments representing the phases linked together by permeable membranes and connected by a peristaltic valve can simulate the gastrointestinal tract of newborns. Presence of human microbiota | Through software, fluid kinetics, transit time, peristaltic movement, temperature, and pH adjustments are regulated | Absence of mucosa The energy density of food is defined by the software | Correlation with human *in vivo* data from volunteers regarding digestion of macromolecules (carbohydrates) and glycemic response. micronutrient digestion | (Minekus, 2015; Bellmann et al. 2016) |

| Model | Simulated phase | Particularities | Benefits | Disadvantages | *In vivo* validation | |
|---|---|---|---|---|---|---|
| DMG | Stomach | System wrapped in a water jacket at 37 °C. Peristaltic contractions. Gastric fluid control, pH, and acid addition control mechanism | It operates with real meals, and peristaltic movements with their implications mimicking the human system. It can be used for fasted or non-fasted conditions. Ease of sample collection. | Simulates only the gastric phase | Correlation with *in vivo* data in human volunteers related to the disintegration of high or low viscosity food particles. | (Wickham et al. 2012) |
| HGS | Stomach | Mechanical process for particle disintegration, with increasing or decreasing intensity. Gastric emptying. Sieve to retain larger particles. | Works with up to 1 liter of the meal. Allows analyses related to both chemical and physical disintegration of particles. | Simulates only the gastric phase | Boiled white rice was offered to piglets and the digestion process followed for 8 hours | (Kong, Singh 2010); Ferrua, Singh, 2015) |
| DIDGI [R] | Stomach and small intestine | Compartments wrapped in the water jacket. Temperature 39 °C. Presence of pH, temperature, and redox sensors. Simulation of anaerobic conditions using nitrogen | Capable of processing complete meals of up to 200 grams. The transparency of the compartments makes it easy to follow the phases | Need for studies with a greater variety of *in vivo food matrices* for validation. Do not simulate absorption but basic agitation only. | *in vitro* and *in vivo* tests, infant formula was offered to 18 piglets and analyzed for lipid and protein digestion. | (Ménard et al. 2018; Dupont et al. 2019) |
| ESIN | Stomach and small intestine (duodenum, jejunum, and ileum) | Emptying of solids and liquids and solids. It portrays the main parameters of digestion. | Absorption simulation. Differential emptying. | Does not have a microbiota simulation | *in vivo* data regarding the administration of pharmaceutical drugs by analyzing human blood and saliva samples | (Guerra et al. 2012; Guerra et al. 2016) |

**Table 3.** *In vitro* studies with microencapsulated and free probiotic strains and simulated digestive phases

| Phases of *in vitro* simulation | Probiotic strain | Microencapsulated or free and microcapsule wall material | Storage period and conditions | Probiotic survival at the end of the storage period | Average viability loss after simulation | References |
|---|---|---|---|---|---|---|
| Stomach, small intestine, and large intestine | *Lactiplantibacillus plantarum* 550 | Microencapsulated by: barley β- glucan, gum arabic, whey protein | - | >9 Log $CFU.g^{-1}$ | Free cells total loss<br>Microcapsules 0.84 Log $CFU.g^{-1}$ | (Yuan et al. 2023) |
| Stomach, small intestine, and large intestine | *Lactiplanibacillus plantarum* 550 | Microencapsulated by: pectin soy protein isolate | 4 weeks<br>4°C | - | Free cells total loss<br>Microcapsules 0.4 to 2 Log $CFU.g^{-1}$ | (Hu et al. 2023) |
| Stomach, small intestine, and large intestine | *Lactiplantibacillus plantarum* LP299V® | Microencapsulated by: maltodextrin, sweet potato flour, gelatin, and pectin | 45 days<br>8°C<br>-18°C | 8.38 Log $CFU.g^{-1}$<br>9.10 Log $CFU.g^{-1}$ | Free Cell 6.04Log $CFU.g^{-1}$<br>Microcapsules 1.33 to 1.88 Log $CFU.g^{-1}$ | (Costa et al. 2023) |
| Stomach, small intestine | *Limosylactobacillus fermentum* NCU001464 | Microencapsulated by: gelatin-xylooligosaccharides, and gum arabic | 10 weeks<br>4°C<br>25°C | 7.23–7.88 Log $CFU.g^{-1}$ | Free cells 5.22 Log $CFU.g^{-1}$<br>Microcapsules 2,35 Log $CFU.g^{-1}$ | (Li et al. 2023) |
| Stomach, small intestine, and large intestine by static method | *Lacticaseibacillus rhamnosus* HN001 | Microencapsulated by: gum arabic, agave fructans, maltodextrin, chicory inulin, trehalose | 8 weeks<br>4°C<br>20°C | >6 Log $CFU.g^{-1}$<br>>8 Log $CFU.g^{-}$ | Free cells 6.21 Log $CFU.g^{-1}$<br>Microcapsules between 3.02 and 4.74 Log $CFU.g^{-1}$ | (Barajas-Àlvarez, González-Ávila, and Espinosa-Andrews 2022) |
| Stomach and intestine | *Lacticaseibacillus rhamnosus*, *Bifidobacterium animalis* | Microencapsulated by: alginate, glycerol, inulin, and oligofructose | 8 days<br>5°C | 7.8 Log $CFU.g^{-1}$<br>8 log $CFU.g^{-1}$ | 0.9 to 1.2 Log $CFU.g^{-1}$ | (Alvarez et al. 2021) |

| Phases of *in vitro* simulation | Probiotic strain | Microencapsulated or free and microcapsule wall material | Storage period and conditions | Probiotic survival at the end of the storage period | Average viability loss after simulation | References |
|---|---|---|---|---|---|---|
| Stomach and intestine | *Kluyveromyces marxianus* VM004 | Microencapsulated by: water soluble chitosan and whey protein concentrate | 90 days<br>8°C<br>25°C | 8 Log $CFU.g^{-1}$ | - | (Vanden Braber et al. 2020) |
| Stomach, small intestine, and large intestine | *Lactiplantibacillus plantarum* LP299V | Free | 16 days<br>5°C | >8 Log $CFU.g^{-1}$ | - | (Costa et al. 2020) |
| Stomach, small intestine, and large intestine | *Lacticaseibacillus rhamnosus* GG (LGG) | Free | 42 days<br>8°C | 8.34 Log $CFU.mL^{-1}$<br>7.50 log $CFU.mL^{-1}$ | - | (Montanari et al. 2020) |
| Mouth, stomach, and intestines | *Lactobacillus paracasei* IMPC2.1 | Free | 30 days<br>4°C | 7 Log $CFU.g^{-1}$ | 2 Log $CFU.g^{-1}$ | (Valerio et al. 2020) |
| Stomach and intestine | *Lactiplantibacillus plantarum* PTCC 1058 | Microencapsulated by: alginate, chitosan, and Na-TPP dual ionic crosslinking agent | - | - | Free cells 4.7 Log $CFU.g^{-1}$ or Ml Microcapsules between 1.58 to 2.05 Log $CFU.g^{-1}$ or mL | (Vaziri, Alemzadeh, and Vossoughi 2018) |
| Stomach and intestine | *Lactiplantibacillus plantarum* CECT 220 and *Lacticaseibacillus casei* CECT 475 | Microparticles based on soy protein, calcium salts | 158 days<br>25°C<br>60% humidity | - | Free cells > 4 Log $CFU.g^{-1}$ or mL Microcapsules between 1 to 2 Log $CFU.g^{-1}$ or mL | (González-Ferrero, Irache, and González-Navarro 2018) |

To correlate the *in vitro* assay with the *in vivo* one, *Lactobacillus casei* was microencapsulated in sodium caseinate and fat matrix. The microcapsules and free cells of the microorganism were submitted to *in vitro* gastrointestinal simulation. Microencapsulated cells had almost 100% survival, a rate well above free cells. To verify these effects *in vivo*, free and microencapsulated probiotic cells were offered via gavage to mice that were sacrificed 5 minutes, 1 hour, and 3 hours after feeding to study the stability of the microcapsules and at 3 and 24 hours to verify the protection imposed by microcapsules to probiotics. The stomach and small intestine were analyzed. The stability study showed that microcapsules were detected in the stomach shortly after administration, there were some after 1 hour and none visible after 3 hours. The difference is that some were detected in the small intestine, and the others are supposed to have been digested in the stomach. The burst test showed that at 3 hours, 10% of the microcapsules were already released and most of the microencapsulated bacteria had already been released at 24 hours. According to the authors, the *in vivo* study with the microcapsules did not correspond to the *in vitro* study (Würth et al. 2015).

Knowing the benefits of probiotic *Escherichia coli* Nissle for inflammatory diseases and its infeasibility caused by the gastric system, its microencapsulation was made for application *in vivo* assays. The free cells of *Escherichia coli* Nissle and the microcapsules formed by alginate, chitosan and cross-linked with $CaCl_2$, passed the *in vitro* gastrointestinal simulation test. Free cells did not survive after 4 hours. The microencapsulated cells had a viability of 46.33%, and live *Escherichia coli* Nissle cells were also identified after 12 hours of *in vitro* digestion. To correlate the results, free and microencapsulated cells were administered to rats with induced colitis. At 8 days after colitis induction, the rats were sacrificed and the colons were analyzed. The *in vivo* assay comparing control with untreated colitis rats and those treated with the microcapsules showed that those treated with the microcapsules recovered body weight and had the activity-reduced disease. The prognostic indicator of the disease showed rates 2.7 times lower for mice treated with the microcapsules than for the other treatments (Luo et al. 2020) positively correlating with the *in vitro* assay.

Many advances have been achieved with simulated studies related to the gastric system, but new research must continue to be carried out to achieve a model with greater similarity to the real one.

## *In Vitro* Studies on Probiotics Administration

Functional foods are increasingly demanded by consumers, among the most popular are those containing probiotic microorganisms. But these beneficial bacteria face many challenges until they reach the colon tissues (Fiocco et al. 2019), so research continues to be carried out either with free or microencapsulated cells. Some of these surveys are shown in Table 3.

## Conclusion

Thanks to the advances achieved by simulation studies, science has found different solutions to public health, nutrition, and food safety problems, in addition to new pharmaceutical formulations being discovered and produced. Health care has been intensified and the use of functional foods containing probiotics is a global megatrend.

Microencapsulation of probiotic bacteria is the most efficient and safe means of transporting these microorganisms through the adverse gastric system to their ideal site of action. In addition, breaking the microcapsules containing the probiotic microorganisms in the core is a key factor in choosing the appropriate technique and the wall materials used. Breaking at the site of action and releasing the nucleus with the minimum necessary number of viable probiotic cells are the ideal characteristics of an efficient microcapsule capable of protecting probiotic microorganisms.

## References

Adebayo-Tayo, B. C., Oluwabunmi, F., Ebenezer, G. Production and evaluation of probioticated mango juice using *Pediococcus pentosaceus* and *Pediococcus acidilactici* during storage at different temperatures. *Journal of Agriculture and Food Research*, vol. 6, p. 100202, 2021. https:// doi: 10.1016/j.jafr.2021.100202

Abbeele, Pieter Van Den, Clara Belzer, Margot Goossens, Michiel Kleerebezem, Willem M. De Vos, Olivier Thas, Rosemarie De Weirdt, Frederiek Maarten Kerckhof, and Tom Van De Wiele. 2013. "Butyrate-Producing Clostridium Cluster XIVa Species Specifically Colonize Mucins in an in Vitro Gut Model." *ISME Journal* 7 (5): 949–61. https://doi.org/10.1038/ismej.2012.158

Abbeele, Pieter Van Den, Charlotte Grootaert, Massimo Marzorati, Sam Possemiers, Willy Verstraete, Philippe Gérard, Sylvie Rabot, Aurélia Bruneau, Sahar El Aidy, Muriel Derrien, Erwin Zoetendal, Michiel Kleerebezem, Hauke Smidt, Tom Van de Wiele.

2010. "Microbial Community Development in a Dynamic Gut Model Is Reproducible, Colon Region Specific, and Selective for Bacteroidetes and Clostridium Cluster IX." *Applied and Environmental Microbiology* 76 (15): 5237–46. https://doi.org/10.1128/AEM.00759-10

Abbeele, Pieter Van den, Stefan Roos, Venessa Eeckhaut, Donald A. Mackenzie, Melanie Derde, Willy Verstraete, Massimo Marzorati, Sam Possemiers, Barbara Vanhoecke, Filip Van Immerseel, Tom Van de Wiele. 2012. "Incorporating a Mucosal Environment in a Dynamic Gut Model Results in a More Representative Colonization by *Lactobacilli*." *Microbial Biotechnology* 5 (1): 106–15. https://doi.org/10.1111/j.1751-7915.2011.00308.x

Alvarez, María Victoria, María Florencia Bambace, Gabriel Quintana, Andrea Gomez-Zavaglia, and María del Rosario Moreira. 2021. "Prebiotic-Alginate Edible Coating on Fresh-Cut Apple as a New Carrier for Probiotic Lactobacilli and Bifidobacteria." *Lwt* 137 (May 2020): 110483. https://doi.org/10.1016/j.lwt.2020.110483

Bagga, Deepika, Johanna Louise Reichert, Karl Koschutnig, Christoph Stefan Aigner, Peter Holzer, Kaisa Koskinen, Christine Moissl-Eichinger, and Veronika Schöpf. 2018. "Probiotics Drive Gut Microbiome Triggering Emotional Brain Signatures." *Gut Microbes* 9 (6): 486–96. https://doi.org/10.1080/19490976.2018.1460015

Baião, Rita, Liliana P. Capitão, Cameron Higgins, Michael Browning, Catherine J. Harmer, and Philip W. J. Burnet. 2022. "Multispecies Probiotic Administration Reduces Emotional Salience and Improves Mood in Subjects with Moderate Depression: A Placebo-Controlled Study." *Psychological Medicine*, 1–11. https://doi.org/10.1017/S003329172100550X

Barajas-Álvarez, Paloma, Marisela González-Ávila, and Hugo Espinosa-Andrews. 2021. "Recent Advances in Probiotic Encapsulation to Improve Viability under Storage and Gastrointestinal Conditions and Their Impact on Functional Food Formulation." *Food Reviews International*, 1–22. https://doi.org/10.1080/87559129.2021.1928691

Barajas-Álvarez, Paloma, Marisela González-Ávila, and Hugo Espinosa-Andrews. 2022. "Microencapsulation of *Lactobacillus Rhamnosus* HN001 by Spray Drying and Its Evaluation under Gastrointestinal and Storage Conditions." *Lwt* 153: 112485. https://doi.org/10.1016/j.lwt.2021.112485

Bellis, Palmira de, Angelo Sisto, and Paola Lavermicocca. 2021. "Probiotic Bacteria and Plant-Based Matrices : An Association with Improved Health-Promoting https://doi.org/10.1016/j.jff.2021.104821

Bellmann, Susann, Jan Lelieveld, Tom Gorissen, Mans Minekus, and Robert Havenaar. 2016. "Development of an Advanced *in Vitro* Model of the Stomach and Its Evaluation versus Human Gastric Physiology." *Food Research International* 88: 191–98. https://doi.org/10.1016/j.foodres.2016.01.030

Bohn, Torsten, Frederic Carriere, Li Day, Amelie Deglaire, Charlotte Egger, Daniela Freitas, Matt Golding, Steven Le Feunteun, A. Macierzanka, O. Menard, B. Miralles, A. Moscovici, R Portmann, I. Recio, D. Rémond, V. Santé-Lhoutelier, T. J. Wooster, U. Lesmes, A. R. Mackie, D. Dupont. 2018. "Correlation between *in Vitro* and in Vivo Data on Food Digestion. What Can We Predict with Static *in Vitro* Digestion Models?" *Critical Reviews in Food Science and Nutrition* 58 (13): 2239–61

Braber, N. L. Vanden, L. I. Díaz Vergara, Y. E. Rossi, C. A. Aminahuel, A. N. Mauri, L. R. Cavaglieri, and M. A. Montenegro. 2020. "Effect of Microencapsulation in Whey Protein and Water-Soluble Chitosan Derivative on the Viability of the Probiotic Kluyveromyces Marxianus VM004 during Storage and in Simulated Gastrointestinal Conditions." *Lwt* 118: 108844. https://doi.org/10.1016/j.lwt.2019.108844

Brodkorb, André, Lotti Egger, Marie Alminger, Paula Alvito, Ricardo Assunção, Simon Ballance, Torsten Bohn, Claire Bourlieu-Lacanal, Rachel Boutrou, Frédéric Carrière, Alfonso Clemente, Milena Corredig, Didier Dupont, Claire Dufour, Cathrina Edwards, Matt Golding, Sibel Karakaya, Bente Kirkhus, Steven Le Feunteun, Uri Lesmes, Adam Macierzanka, Alan R. Mackie, Carla Martins, Sébastien Marze, David Julian McClements, Olivia Ménard, Mans Minekus, Reto Portmann, Cláudia N. Santos, Isabelle Souchon, R. Paul Singh, Gerd E. Vegarud, Martin S. J. Wickham, Werner Weitschies & Isidra Recio. 2019. "INFOGEST Static in Vitro Simulation of Gastrointestinal Food Digestion." *Nature Protocols* 14 (4): 991–1014. https://doi.org/10.1038/s41596-018-0119-1

Collen, Alanna. 2016. 10% Human: How Your Body's Microbes Hold the Key to Health and Happiness. Editora Harper Paperbacks

Costa, Nataly A., Laura R. Silveira, Ester P. Amaral, Gabriel C. Pereira, Daniele A. Paula, Érica Nascif R. Vieira, Eliane M. F. Martins, P. C. Strigheta, Bruno R. C. Leite Junior, and Afonso M. Ramos. 2023. "Use of Maltodextrin, Sweet Potato Flour, Pectin and Gelatin as Wall Material for Microencapsulating *Lactiplantibacillus Plantarum* by Spray Drying : Thermal Resistance, in Vitro Release Behavior, Storage Stability, and Physicochemical Properties." *Food Research International* 164: 112367. https://doi.org/10.1016/j.foodres.2022.112367

Costa, Nataly Almeida, Daniele Almeida Paula, Juliana Dias Brêda, Érica Nascif Rufino Vieira, Eliane Maurício F. Martins, and Afonso Mota Ramos. 2020. "A Symbiotic Dessert Composed of Yam (*Dioscorea* Sp.) and Ubá Mango Pulp (Mangifera Indica L.)." *LWT* 133 (11074). https://doi.org/10.1016/j.lwt.2020.110074

Derrien, Muriel, and Johan E.T. Hylckama Vlieg. 2015. "Fate, Activity, and Impact of Ingested Bacteria within the Human Gut Microbiota." *Trends in Microbiology* 23 (6): 354–66. https://doi.org/10.1016/j.tim.2015.03.002.

Dupont, D., M. Alric, S. Blanquest, G Bornhorst, C. Cueva, A. Deglaire, S. Denis, M Ferrua, R Havenaar, J Lelieveld, A R Mackie, M Marzorati, O Menard, M Minekus, B Miralles, I Recio, P Van den Abbeele. 2019. "Can Dynamic in Vitro Digestion Systems Mimic the Physiological Reality?" *Food Science and Nutrition* 59 (10): 1546–62. https://doi.org/10.1080/10408398.2017.1421900

Ferrua, Maria J., and R. Paul Singh. 2015. "Human Gastric Simulator (Riddet Model)." *The Impact of Food Bioactives on Health: In Vitro and Ex Vivo Models*, 1–327. https://doi.org/10.1007/978-3-319-16104-4

Fioravanti, Maria I. A., and Marcelo Antônio Morgano. 2021. "Uma Abordagem Dos Ensaios in Vitro Para Estimar a Absorção Dos Minerais Em Fórmulas Infantis." *Brazilian Journal of Food Technology* 24: 1–12. https://doi.org/10.1590/1981-6723.09820

González-Ferrero, C., J. M. Irache, and C. J. González-Navarro. 2018. "Soybean Protein-Based Microparticles for Oral Delivery of Probiotics with Improved Stability during Storage and Gut Resistance." *Food Chemistry* 239: 879–88.

Guerra, Aurélie, Sylvain Denis, Olivier le Goff, Vincent Sicardi, Olivier François, Anne Françoise Yao, Ghislain Garrait, Aimé Pacifique Manzi, Eric Beyssac, Monique Alric, Stéphanie Blanquet-Diot. 2016. "Development and Validation of a New Dynamic Computer-Controlled Model of the Human Stomach and Small Intestine." *Biotechnology and Bioengineering* 113 (6): 1325–35. https://doi.org/10.1002/bit.25890

Guerra, Aurélie, Lucie Etienne-Mesmin, Valérie Livrelli, Sylvain Denis, Stéphanie Blanquet-Diot, and Monique Alric. 2012. "Relevance and Challenges in Modeling Human Gastric and Small Intestinal Digestion." *Trends in Biotechnology* 30 (11): 591–600. https://doi.org/10.1016/j.tibtech.2012.08.001

Hu, Ronghai, Dejun Dong, Jielun Hu, and Huan Liu. 2023. "Improved Viability of Probiotics Encapsulated in Soybean Protein Isolate Matrix Microcapsules by Coacervation and Cross-Linking Modification." *Food Hydrocolloids* 138: 108457. https://doi.org/10.1016/j.foodhyd.2023.108457

Hur, Sun Jin, Beong Ou Lim, Eric A. Decker, and D. Julian McClements. 2011. "In Vitro Human Digestion Models for Food Applications." *Food Chemistry* 125 (1): 1–12. https://doi.org/10.1016/j.foodchem.2010.08.036

Langdon, Amy, Nathan Crook, and Gautam Dantas. 2016. "The Effects of Antibiotics on the Microbiome Throughout Development and Alternative Approaches for Therapeutic Modulation." *Genome Medicine* 8 (39). https://doi.org/10.1186/s13073-016-0294-z

Li, Cheng, Yiming Hu, Songnan Li, Xueer Yi, Shuaibo Shao, Wenwen Yu, and Enpeng Li. 2023. "Biological Factors Controlling Starch Digestibility in Human Digestive System." *Food Science and Human Wellness* 12 (2): 351–58. https://doi.org/10.1016/j.fshw.2022.07.037

Li, Huiyu, Fei Peng, Jia Xin Lin, Tao Xiong, and Tao Huang. 2023. "Preparation of Probiotic Microcapsules Using Gelatin-Xylooligosaccharides Conjugates by Spray Drying: Physicochemical Properties, Survival, Digestion Resistance and Colonization." *Food Bioscience* 52 (235): 102462. https://doi.org/10.1016/j.fbio.2023.102462

Lillo-Pérez, Siegried, María Guerra-Valle, Patricio Orellana-Palma, and Guillermo Petzold. 2021. "Probiotics in Fruit and Vegetable Matrices: Opportunities for Nondairy Consumers." *LWT* 151 (112106). https://doi.org/10.1016/J.LWT.2021.112106

Kong, F.; Singh R. P. A human gastric simulator (HGS) to study food digestion in the human stomach. *Journal of Food Science* 75(9):E627–E635, 2010. doi: 10.1111/j.1750-3841.2010.01856.x

Mao, Like, Qiuyue Pan, Fang Yuan, and Yanxiang Gao. 2019. "Formation of Soy Protein Isolate-Carrageenan Complex Coacervates for Improved Viability of Bifidobacterium Longum during Pasteurization and in Vitro Digestion." *Food Chemistry* 276: 307–14. https://doi.org/10.1016/j.foodchem.2018.10.026

Marze, S. 2017. "Bioavailability of Nutrients and Micronutrients : Advances in Modeling and *In Vitro* Approaches." *Annual Rev. Food Sci. Technol* 8: 35–55. https://doi.org/10.1146/annurev-food-030216-030055

Ménard, O., C. Bourlieu, S. C. De Oliveira, N. Dellarosa, L. Laghi, F. Carrière, F. Capozzi, D. Dupont, and A. Deglaire. 2018. "A First Step towards a Consensus Static in Vitro Model for Simulating Full-Term Infant Digestion." *Food Chemistry* 240: 338–45. https://doi.org/10.1016/j.foodchem.2017.07.145

Minekus, M. 2015. "The TNO Gastro-Intestinal Model (TIM)." In *The Impact of Food Bioactives on Health: In Vitro and Ex Vivo Models*, 37–46. Utrechtseweg: TNO Triskelion. https://doi.org/10.1007/978-3-319-16104-4-5

Minekus, M, M. Alminger, P. Alvito, S. Ballance, T. Bohn, C. Borlieu, F. Carrière, R. Boutrou, M. Corredig, D. Dupont, C. Dufour, L. Egger, M. Golding, S. Karakaya, B. Kirkhus, S. Le Feunteun, U. Lesmes, A. Macierzanka, A. Mackie, S. Marze, D. J. McClements, O. Ménard, I. Recio, C. N. Santos, R. P. Singh, G. E. Vegarud, M. S. J. Wickham, W. Weitschies and A. Brodkorb. 2014. "A Standardised Static in Vitro Digestion Method Suitable for Food – an International Consensus." *Food & Function* 5: 1113–24. https://doi.org/10.1039/c3fo60702j

Montanari, Shimeny Ramos, Bruno Ricardo de Castro Leite Júnior, Maurilio Lopes Martins, Afonso Mota Ramos, Mirella Lima Binoti, Renata Cristina de Almeida Bianchini Campos, André Narvaes da Rocha Campos, and Eliane Maurício Furtado Martins. 2020. "*In Vitro* Gastrointestinal Digestion of a Peanut, Soybean, Guava, and Beet Beverage Supplemented with *Lactobacillus Rhamnosus* GG." *Food Bioscience* 36 (October 2018): 100623. https://doi.org/10.1016/j.fbio.2020.100623

Ningtyas, Dian Widya, Bhesh Bhandari, Nidhi Bansal, and Sangeeta Prakash. 2019. "The Viability of Probiotic *Lactobacillus Rhamnosus* (Non-Encapsulated and Encapsulated) in Functional Reduced-Fat Cream Cheese and Its Textural Properties during Storage." *Food Control* 100: 8–16. https://doi.org/10.1016/j.foodcont.2018.12.048.

Pavli, Foteini, Ioanna Kovaiou, Georgia Apostolakopoulou, Anastasia Kapetanakou, Panagiotis Skandamis, George John E. Nychas, Chrysoula Tassou, and Nikos Chorianopoulos. 2017. "Alginate-Based Edible Films Delivering Probiotic Bacteria to Sliced Ham Pretreated with High-Pressure Processing." *International Journal of Molecular Sciences* 18 (9): 2–16. https://doi.org/10.3390/ijms18091867

Prisco, Annachiara De, and Gianluigi Mauriello. 2016. "Probiotication of Foods : A Focus on Microencapsulation Tool." *Trends in Food Science & Technology* 48: 27–39. https://doi.org/10.1016/j.tifs.2015.11.009

Rasika, Dissanayake, Janak K. Vidanarachchi, Ramon Silva Rocha, Celso F. Balthazar, Adriano G. Cruz, Anderson S. Sant Ana, and Chaminda Senaka Ranadheera. 2021. "Plant-Based Milk Substitutes as Emerging Probiotic Carriers," 8–20. https://doi.org/10.1016/j.cofs.2020.10.025

Saladin, Kenneth S. 2016. *Human Anatomy. Nature*. 2nd ed. Vol. 261. New York, NY 10020: McGraw-Hil. https://doi.org/10.1038/261010c0

Terpou, Antonia, Aikaterini Papadaki, Iliada K Lappa, Vasiliki Kachrimanidou, Loulouda A Bosnea, and Nikolaos Kopsahelis. 2019. "Probiotics in Food Systems: Significance and Emerging Strategies Towards Improved Viability and Delivery of Enhanced

Beneficial Value." *Nutrients* 11 (7): 1–32. https://www.mdpi.com/2072-6643/11/7/1591

Thorakkattu, Priyamvada, Anandu C. Khanashyam, Kartik Shah, Karthik Sajith Babu, Anjaly Shanker Mundanat, Aiswariya Deliephan, Gitanjali S. Deokar, Chalat Santivarangkna, and Nilesh Prakash Nirmal. 2022. "Postbiotics: Current Trends in Food and Pharmaceutical Industry." *Foods* 11 (19): 1–29. https://doi.org/10.3390/foods11193094

Valerio, Francesca, Maria Grazia, Gabriella Santagata, Floriana Boscaino, Costantina Barbarisi, Mariaelena Di, Anna Rita, Stella Lisa, and Paola Lavermicocca. 2020. "The Viability of Probiotic *Lactobacillus Paracasei* IMPC2. 1 Coating on Apple Slices during Dehydration and Simulated Gastro-Intestinal Digestion." *Food Bioscience* 34 (100533). https://doi.org/10.1016/j.fbio.2020.100533

Vaziri, Asma Sadat, Iran Alemzadeh, and Manouchehr Vossoughi. 2018. "Improving Survivability of *Lactobacillus Plantarum* in Alginate-Chitosan Beads Reinforced by Na-Tripolyphosphate Dual Cross-Linking." *Lwt* 97: 440–47. https://doi.org/10.1016/j.lwt.2018.07.037

Wickham, M. J. S., R. M. Faulks, J. Mann, and G. Mandalari. 2012. "The Design, Operation, and Application of a Dynamic Gastric Model." *Dissolution Technologies* 19 (3): 15–22. https://doi.org/10.14227/DT190312P15

Würth, Rebecca, Gabriele Hörmannsperger, Julia Wilke, Petra Foerst, Dirk Haller, and Ulrich Kulozik. 2015. "Protective Effect of Milk Protein Based Microencapsulation on Bacterial Survival in Simulated Gastric Juice versus the Murine Gastrointestinal System." *Journal of Functional Foods* 15: 116–25. https://doi.org/10.1016/j.jff.2015.02.046

Yuan, Cong, Ronghai Hu, Li He, Jielun Hu, and Huan Liu. 2023. "Extraction and Prebiotic Potential of β-Glucan from Highland Barley and Its Application in Probiotic Microcapsules." *Food Hydrocolloids* 139: 108520. https://doi.org/10.1016/j.foodhyd.2023.10852

# About the Editors

**Daniele De Almeida Paula**
Federal Institute of Paraná, IFPR,
Campus Assis Chateubriand, Paraná, Brazil

**Professor Dr. Daniele de Almeida Paula** received her master´s and doctoral degrees in food science and technology from the Federal University of Viçosa – UFV in 2019. She is currently working as a prof. at the Federal Institute of Paraná IFPR, campus Assis Chateaubriand. She works in research areas involving the use of microencapsulation techniques of probiotic microorganisms, application of probiotics in the vegetable matrix, and viability studies of these microorganisms when subjected to gastrointestinal conditions simulated *in vitro*. She has projects, publications, and book chapters involving his area of expertise. In addition to the areas mentioned, it has also started research focused on the development of probiotic products using Kefir and Kombucha.

**Karen Cristina Guedes Silva**
Professor at the Federal Institute of Paraná - IFPR, Campus Assis Chateaubriand, Paraná, Brazil

**Dr. Karen Cristina Guedes Silva** works as a professor at the Federal Institute of Paraná - IFPR, Campus Assis Chateaubriand (Assis Chateaubriand/PR, Brazil), teaching basic, technical, and technological education. Professor Karen Silva received her master´s and doctoral degrees in Food Engineering from the University of Campinas - UNICAMP (Campinas/SP, Brazil), with a doctoral period in partnership with the International Iberian Nanotechnology Laboratory – INL (Braga, Portugal). She holds a degree in Mathematics Education and a Bachelor's degree in Food Engineering from the Federal University of Lavras - UFLA (Lavras/MG, Brazil), and during her undergraduate studies, she spent a year at the University of Santiago de Compostela – USC (Lugo, Spain). She has experience in the encapsulation of bioactive compounds for food application, an area that encompasses a significant portion of her publications, such as biopolymer blends for encapsulation systems, evaluation of protection and controlled delivery of bioactive through in vitro digestibility, and encapsulation of symbiotics and probiotics.

**Nataly de Almeida Costa**
Federal University of Viçosa, Minas Gerais, Brazil

Master in Food Science and Technology from the Federal University of Viçosa where he developed a symbiotic dessert based on vegetables. Bachelor in Food Science and Technology from the Federal Institute of Education, Science, and Technology of Southeast Minas Gerais Campus Rio Pomba. Currently, he is a PhD student in Food Science and Technology at the Federal University of Viçosa (DTA-UFV) where he is part of the Laboratory of Innovation in Food Processing - LIPA/UFV and has been developing projects in the area of functional foods, use of techniques of microencapsulation of probiotic microorganisms, application of probiotics in plant matrix and the survival of these microorganisms when subjected to gastrointestinal conditions.

# Index

## M



## N



## O



## P

## R



## S